Collins

ROYAL
OBSERVATORY
GREENWICH

Night Sky
ALMANAC
A STARGAZER'S
GUIDE TO

2026

Radmila Topalovic &
Dominic Ford

Published by Collins
An imprint of HarperCollins Publishers
1 Robroyston Gate
Glasgow G33 1JN
collins.co.uk

HarperCollins Publishers
Macken House, 39/40 Mayor Street Upper
Dublin 1
D01 C9W8
Ireland

In association with Royal Museums Greenwich, the group name for the National Maritime
Museum, Royal Observatory Greenwich, the Queen's House and *Cutty Sark*
www.rmg.co.uk

First published 2025

© HarperCollins Publishers 2025
Text © Radmila Topalovic
Diagrams © Dominic Ford
Cover illustrations © Julia Murray
Images and illustrations © see acknowledgements page 267

A catalogue record for this book is available from the British Library

ISBN 978-0-00-874782-4

10 9 8 7 6 5 4 3 2 1

Printed in the UK using 100% Renewable Electricity at CPI Group (UK) Ltd

If you would like to comment on any aspect of this book, please contact us at the above address
or online. e-mail: collins.reference@harpercollins.co.uk

Contents

Data used in the Night Sky Almanac

The data given in this Almanac, such as timings and distances between objects, have been determined by the following sources: the Astronomical Applications Department of the US Naval Observatory; the Sky Events Calendar by Fred Espenak and Sumit Dutta (NASA's Goddard Space Flight Center); the NASA Jet Propulsion Laboratory (JPL) Horizons System; the International Astronomical Union's (IAU) Minor Planet Center; In-The-Sky.org.

Introduction

The aim of this book is to help people find their way around the night sky at any time of the year, by showing how the stars that are visible change from month to month and by highlighting various events that occur during 2026. The objects and events described may be observed with the naked eye, or nothing more complicated than a pair of binoculars.

The charts in the book have been specifically designed for use anywhere in the world. A full description of how to use and understand the monthly charts is given on pages 40–43.

Sunrise, sunset and twilight

The conditions for observing naturally vary over the course of the year and one's location on Earth. Sunrise and sunset vary considerably, depending in particular on one's latitude. Sunrise and sunset times are given each month for nine different locations around the world. These places are shown in **bold** typeface on the world map on pages 44–45. Sunrise and sunset times are given for the first and last days in every month, for these locations. Another factor that influences what may be seen is twilight at dusk and dawn. The diagrams on pages 252–255 show how this varies for the nine locations, which have been chosen to show the range of variation. The different stages of twilight and how they affect observing are also explained there.

Moonlight

Moonlight will affect the visibility of objects, providing a natural form of light pollution. At Full Moon, it may be very difficult to see some of the fainter stars and objects, and even when the Moon is at a smaller phase it may seriously interfere with visibility if it is near the stars or planets in which you are interested. A full lunar calendar is given for each month and may be used to plan night sky observations. It may also be useful to look at moonrise and moonset times for your particular location.

The celestial sphere

All the objects in the sky (including the Sun, Moon and stars) appear to lie at some indeterminate distance on a large sphere, centred on the Earth. This *celestial sphere* has various reference

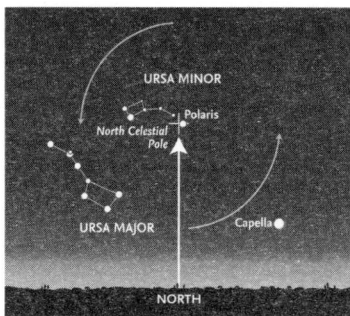

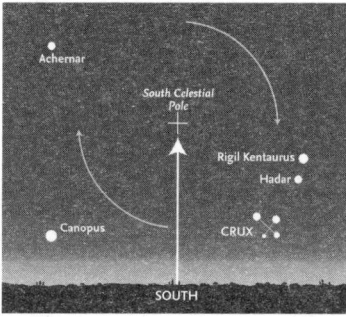

The altitude of the Celestial Pole equals the observer's latitude.

points and features that are related to those of the Earth. If the Earth's rotational axis (an imaginary line through the North and South Poles and the core of the Earth) is extended, for example, it points to the North and South Celestial Poles. Similarly, the **celestial equator** lies in the same plane as the Earth's equator, it divides the sky into northern and southern hemispheres.

It is useful to know some of the astronomy terms for various parts of the sky. As seen by an observer, half of the celestial sphere is invisible at any point in time; these objects will be below the horizon. The point directly overhead is known as the **zenith**, and the (invisible) one below one's feet as the **nadir**. The line running from the north point on the horizon, up through the zenith and then down to the south point is the **meridian**. This is an important invisible line in the sky, because objects are highest in the sky, and thus easiest to see, when they cross the meridian in the south. Objects are said to **transit** when they cross this line in the sky.

In this book, reference is frequently made in the text and in the diagrams to the standard compass points around the horizon. The position of any object in the sky specific to the

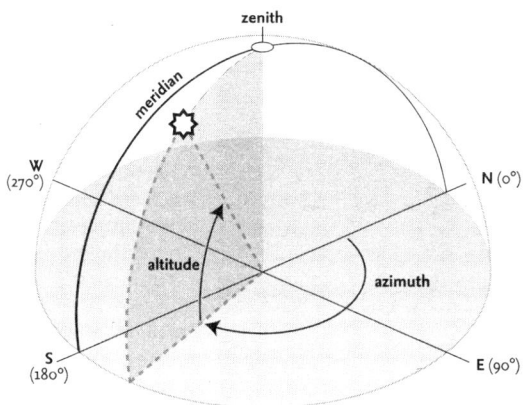

Measuring altitude and azimuth on the celestial sphere.

observer's location on Earth may be described by its ***altitude*** (measured in degrees above the horizon) and its ***azimuth*** (measured in degrees from north 0°, through east 90°, south 180° and west 270°).

The celestial sphere appears to rotate about an invisible axis, running between the North and South Celestial Poles. The location (i.e. the altitude) of the Celestial Poles depends entirely on the observer's latitude on Earth. At the North Pole (latitude 90°), the North Celestial Pole (NCP) would be directly overhead (at the zenith, or at an altitude of 90°).

Right ascension and declination

Astronomers use a more precise method to measure the positions of objects, which does not depend on the observer's position on Earth (and thus on their local horizon). This involves the two coordinates, ***right ascension*** (RA) and ***declination*** (dec). Right ascension is measured eastwards (to the left) from the ***First Point of Aries*** (page 84) in hours and minutes of time (and very occasionally, in seconds). The sky appears to rotate by

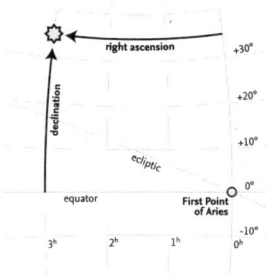

Measuring right ascension (RA) and declination (dec) on the celestial sphere.

15 angular degrees in one hour, one full rotation of 360° takes 24 hours.

All objects in the sky appear to be located on an imaginary sphere: the celestial sphere. There are, however, certain fixed points on the celestial sphere, related to points on the Earth. The North Celestial Pole (NCP) and the South Celestial Pole (SCP) are located in line with the projection of the Earth's rotational axis onto that sphere. In the north, the NCP is very close to Polaris, which has been known as the North Star since antiquity. In a similar way, the celestial equator is the projection onto the sphere of the Earth's equator. The second coordinate, declination, is simply the angular distance, in degrees, north or south of the celestial equator. The Sun has a declination of zero when it appears to cross the celestial equator at the equinoxes. The NCP marks a declination of +90° or 90°N, the SCP conversely is at declination −90° or 90°S.

The ecliptic and the zodiac
Another important line on the celestial sphere is the Sun's apparent path against the background stars – in reality the result of the Earth's orbit around the Sun. This is known as the *ecliptic*. The point where the Sun, apparently moving along the ecliptic, crosses the celestial equator from south to north is known as the vernal (or spring) equinox, which occurs on 20 March. At this time (and at the autumnal equinox, on 22 or 23 September, when the Sun crosses the celestial equator from north to south) day and night are almost exactly equal in length. The vernal equinox is currently located in the constellation of Pisces, meaning the Sun is moving through Pisces on this day. It is important in astronomy because it defines the zero point

for the system of celestial coordinates (right ascension) not used in book.

The Moon and planets are to be found in a band of sky all the way around the celestial sphere that extends 8° on either side of the ecliptic. This is because the orbits of the Moon and planets are inclined at various angles to the ecliptic (i.e., to the plane of the Earth's orbit). This band of sky is known as the zodiac and, when originally devised, consisted of twelve **constellations**, all of which were considered to be exactly 30° wide. When the constellation boundaries were formally established by the International Astronomical Union in 1930, adjustments were made resulting in the ecliptic passing through thirteen constellations, and the Moon and planets passing through several other constellations that are adjacent to the original twelve.

The constellations

Since ancient times, the celestial sphere has been divided into various constellations, most dating back to antiquity, and usually associated with certain myths or legendary people and animals. The boundaries of the 88 constellations across the whole celestial sphere have been fixed by international agreement and their names (in Latin) are largely derived from Greek or Roman originals. A full list of all the constellations is given on pages 258–260, with their abbreviations, their genitive forms and English names. Other naming schemes exist for fainter stars but these are not used in this book. Some of the names of the most prominent stars are of Greek or Roman origin, but many are derived from Arabic names. Many bright stars have no individual names and, for many years, stars were identified by terms such as 'the star in Hercules' right foot'. A more sensible scheme was introduced by the German astronomer Johannes Bayer in the early seventeenth century. Following his scheme – which is still used today – most of the brightest stars are identified by a Greek letter followed by the genitive form of the constellation's Latin name. An example is the Pole Star, also known as Polaris and α Ursae Minoris (abbreviated α UMi). The Greek alphabet is shown on page 260.

Asterisms

Apart from the constellations, certain groups of stars, which may form a part of a larger constellation or cross several constellations, are readily recognizable and have been given individual names. These groups are known as **asterisms**, and the most famous (and well-known) is the **Plough** or **Big Dipper**, the common name for the seven brightest stars in the constellation of **Ursa Major**, the Great Bear. The names and details of some asterisms mentioned in this book are given in the list on page 261.

Magnitudes

The brightness of a star, planet or other body is given in magnitudes (mag.). This is a mathematically defined scale where larger (positive) numbers indicate a fainter object. The scale extends beyond the zero point to negative numbers for very bright objects. (Sirius, the brightest star in the night sky is mag. –1.4.) Most observers are able to see stars as faint as about mag. 6, under very clear skies.

The Moon

Although the daily spin of the Earth carries the sky from east to west (objects in the south rise in the east and set in the west) the Moon gradually moves eastwards relative to the background stars by approximately its diameter in an hour; this is equivalent to about half a degree across the sky. Its apparent eastward motion and change in phase are caused by its orbit around the Earth and changing position relative to the Sun.

Normally, in its orbit around the Earth, the Moon passes above or below the direct line between the Earth and the Sun (at New Moon) or outside the area obscured by the Earth's shadow (at Full Moon). Occasionally, however, the three bodies are more or less perfectly aligned to give an **eclipse**: a solar eclipse at New Moon or a lunar eclipse at Full Moon. Depending on the exact circumstances, a solar eclipse may be merely partial (when the Moon does not cover the whole of the Sun's disk); annular (when the Moon is too far from Earth

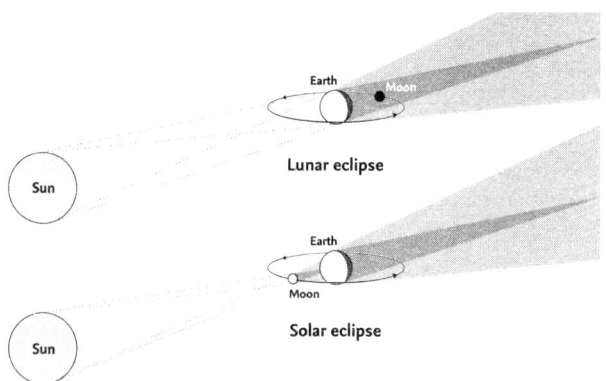

When the Moon passes directly through the Earth's shadow (top), a lunar eclipse occurs. When it passes in front of the Sun (below), a solar eclipse occurs.

in its orbit to appear large enough to hide the whole of the Sun); or total. Total and annular eclipses are visible from very restricted areas of the Earth, but partial eclipses are normally visible over a wider area.

Precautions must always be taken when viewing even partial phases of a solar eclipse to avoid damage to your eyes. Only ever use proper eclipse glasses, or a proper solar filter over the full objective of a telescope. The glass 'solar filters' sometimes provided with cheap telescopes should never be used, they are unsafe.

Somewhat similarly, at a lunar eclipse, the Moon may pass through the lighter outer zone of the Earth's shadow, the *penumbra*, called a penumbral eclipse, which is not generally perceptible to the naked eye; part of the Moon could pass within the darkest part of the Earth's shadow, the umbra, in a partial eclipse; or it could move completely within the *umbra*, in a total eclipse. Unlike solar eclipses, lunar eclipses are visible from large areas of the Earth.

Occasionally, as it moves across the sky, the Moon passes between the Earth and individual planets or distant stars, giving rise to an **occultation**. As with solar eclipses, such occultations are visible from restricted areas of the world.

Angular distance can be measured approximately by holding one hand at arm's length. The various angles are shown in the diagram, together with the separations of the various stars in the Plough or Big Dipper and also for the stars around the constellation of Orion.

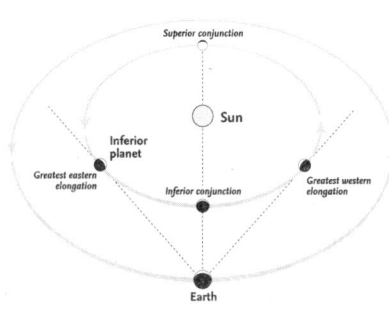

Inferior planet.

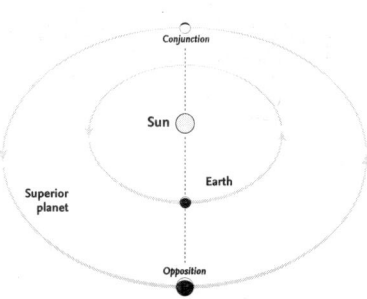

Superior planet.

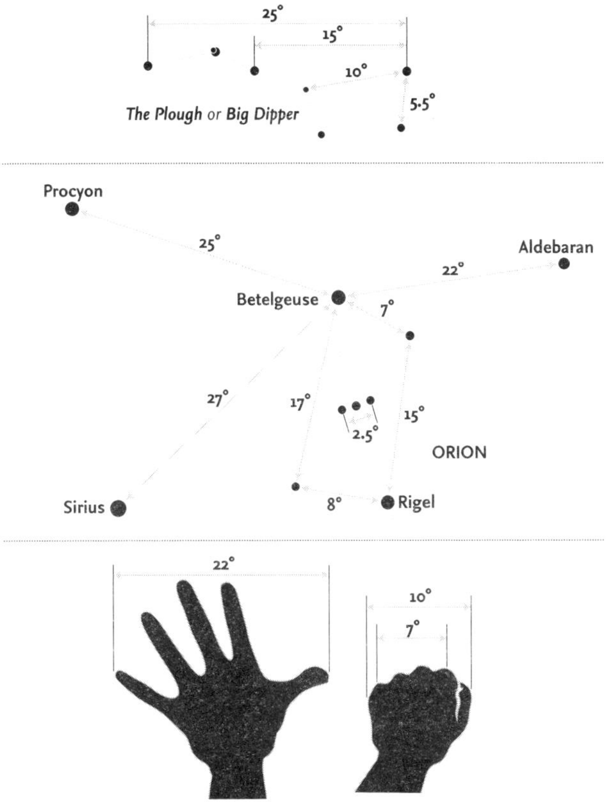

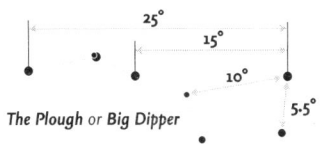

25°

15°

10°

5.5°

The Plough or *Big Dipper*

Procyon

25°

Aldebaran

22°

Betelgeuse

7°

27°

17°

15°

2.5°

ORION

Sirius

8°

Rigel

22°

10°

7°

Measuring angles in the sky.

The planets
The planets are always moving against the background stars,
therefore they are treated in some detail in the monthly pages
and information is given regarding when they are close to the
Sun, other planets, the Moon or any of five bright stars that lie

near the ecliptic. Such events are known as *appulses* or, more frequently, as *conjunctions*. (There are technical differences in the way these terms are defined – and should be used – in astronomy, but these need not concern us here.)

The conditions of most favourable visibility depend on whether the planet is one of the two known as *inferior planets* – planets that orbit the Sun within the Earth's orbit (Mercury and Venus) – or one of the five *superior planets* – planets that orbit the Sun beyond the Earth. Of the latter, three (Mars, Jupiter and Saturn) are covered in detail and these are visible to the naked eye. Occasionally, details of the fainter superior planets, Uranus and Neptune, are included, especially when they come to opposition.

The inferior planets are most readily seen at eastern or western elongation, when their angular distance from the Sun (separation from the Sun in degrees) is greatest. Superior planets are best seen at *opposition*, when they are directly opposite the Sun in the sky and cross the meridian at local midnight.

Events

A number of interesting events are shown in the diagrams for each month. They involve the planets and the Moon, sometimes showing them in relation to specific stars. Events have been chosen as they will appear from one of three different locations: from London; from the central region of the USA; or from Sydney in Australia. Naturally, these events are visible from other locations, but the appearance of the objects on the sky will differ slightly from the diagrams. A list of major astronomical events in 2026 is given on pages 24–25.

Meteors

At some time or other, nearly everyone has seen a *meteor* – a 'shooting star' – as it flashed across the sky. The particles that cause meteors – known technically as 'meteoroids' – normally range in size from that of a grain of sand (or even smaller) to the size of a pea. *Fireballs* or *bolides* are very bright meteors (brighter than mag. −4) that are caused by objects up to 1 metre

Geminid meteor shower, with radiant superimposed.

in size. Fireballs sometimes cause sonic booms that may be heard some time after the meteor is seen. On any night of the year there are occasional meteors, known as sporadics, that may travel in any direction. These occur at a rate that is normally between three and eight per hour.

Far more important, however, are ***meteor showers***, which occur at fixed periods of the year, when the Earth encounters a trail of particles left behind by a comet or, very occasionally, by a minor planet (asteroid). Meteors always appear to diverge from a single point on the sky, known as the radiant, and the radiants of major showers are shown on the charts. Meteors that come from a circular area 8° in diameter around the radiant are classed as belonging to the particular shower. All others that do not come from that area are sporadics (or, occasionally from another shower that is active at the same time). A list of the major meteor showers is given on page 20.

Looking directly at the radiant is not the most effective way of seeing meteors. They are most likely to be noticed if one is looking about 40–45° away from the radiant position. This is approximately two hand-spans as shown in the diagram for measuring angles on page 13.

Motion of the planets

All the outer planets from Mars to Neptune exhibit periods of retrograde motion, when, instead of their normal, orderly progress across the sky from west to east (known as direct motion), they reverse direction and travel from east to west. This retrograde movement continues for a while and then they reverse direction and resume direct motion.

When only the outer planets from Mars to Saturn were known, retrograde motion was one of the main problems that faced astronomers and astrologers when it was believed that the whole Universe was centred on the Earth (in a geocentric universe).

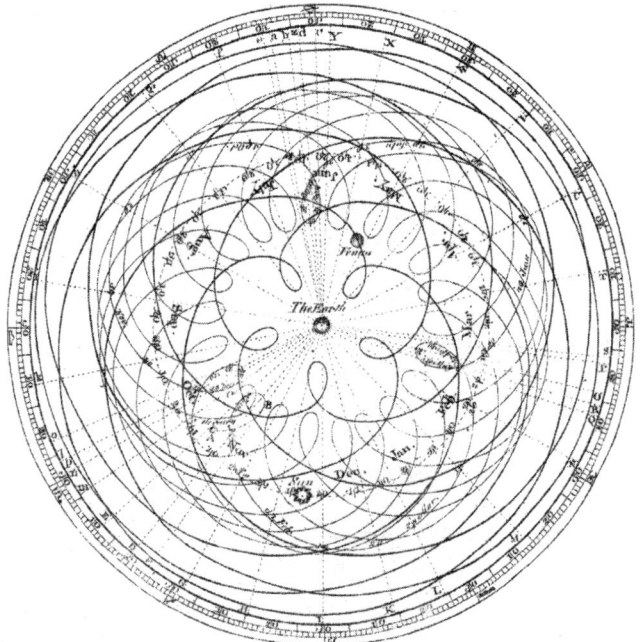

The highly complex pattern of planetary motions that had to be explained on the geocentric model.

The result of this behaviour was a highly complex pattern, which proved difficult to explain on a geocentric model.

Because of the notion that movements in the planetary realm could only occur in 'perfect' circles, the idea was introduced that the planets moved on small circles (epicycles) that were themselves carried around the Earth on circular orbits. This concept was first introduced by Apollonius of Perga, who lived around 240–190 BCE. He studied geometry and astronomy, but most of his writings are lost. The crater Apollonius on the Moon carries his name (see page 19). The various concepts were developed by the Greek astronomer Hipparchus (c.190–120 BCE), who also has a lunar crater named after him (see below).

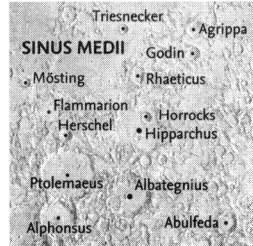

The lunar crater Hipparchus is located near the centre of the Moon.

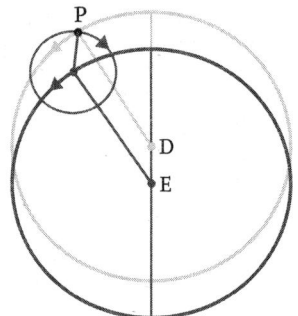

The planet (P) supposedly orbited on a small circle, carried round a larger circle, centred at the deferent (D), offset from the Earth (E).

Apart from 'explaining' the retrograde motion of the planets, the epicyclic theory also provided a solution to the apparent changes in the distances of the planets from the Earth.

This way of explaining the motion of the planets, where the circular epicycle was carried around the Earth in a larger circular orbit, prevailed for some years.

But even this idea proved inadequate to describe the motion of the planets, so the concept of the deferent was added. In this, the circular motion of the epicycle was carried around the Earth on a circle that was itself not centred on the Earth, but offset from its centre.

The use of the epicycle and deferent was developed and propagated by the great astronomer Ptolemy (c.100–170 CE), who found that he had to introduce further terms, which

The astronomer Ptolemy, with tools for measuring the stars.

he denoted the 'eccentric' and the 'equant'. This further complicated the situation. All this complexity became redundant, of course, as soon as the view of Copernicus prevailed, in which the Earth itself orbited the Sun. It became apparent that the planets displayed retrograde motion when the Earth 'caught up' and 'passed' the planets in their orbits.

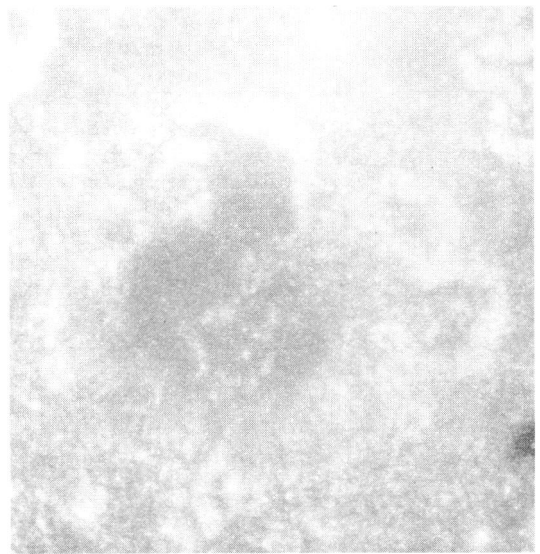

The somewhat indistinct crater Apollonius lies northeast of the Mare Fecunditatis. Its western rim is overlain by two smaller craters, and its flat central floor is flooded with lava.

Meteor Showers

Shower	Dates of activity 2026	Date of maximum	Possible hourly rate
Quadrantids	28 Dec. to 12 Jan.	4 Jan.	120
α-Centaurids	28 Jan. to 21 Feb.	8 Feb.	6
γ-Normids	25 Feb. to 28 Mar.	14 Mar.	6
April Lyrids	16 to 25 Apr.	22 Apr.	18
π-Puppids	15 to 28 Apr.	24 Apr.	var.
η-Aquariids	19 Apr. to 28 May	6 May	40
α-Capricornids	3 Jul. to 15 Aug.	30 Jul.	5
Southern δ-Aquariids	12 Jul. to 23 Aug.	30 Jul.	25
Piscis Austrinids	15 Jul. to 10 Aug.	28 Jul.	5
Perseids	17 Jul. to 24 Aug.	13 Aug.	150
α-Aurigids	28 Aug. to 5 Sept.	1 Sep.	6
Southern Taurids	10 Sep. to 20 Nov.	10 Oct.	5
Orionids	2 Oct. to 7 Nov.	21 Oct.	15
Draconids	6 to 10 Oct.	9 Oct.	var.
Northern Taurids	20 Oct. to 10 Dec.	12 Nov.	5
Leonids	6 to 30 Nov.	18 Nov.	15
Phoenicids	28 Nov. to 9 Dec.	2 Dec.	var.
Puppid-Velids	1 to 15 Dec.	7 Dec.	10
Geminids	4 to 17 Dec.	14 Dec.	120
Ursids	17 to 26 Dec.	22 Dec.	10

Some Interesting Objects

Messier IC / NGC	Name	Type	Constellation
—	47 Tucanae	globular cluster	Tucana
—	Hyades	open cluster	Taurus
—	Double Cluster	open cluster	Perseus
—	Melotte 11	open cluster	Coma Berenices
M3	—	globular cluster	Canes Venatici
M4	—	globular cluster	Scorpius
M8	Lagoon Nebula	gaseous nebula	Sagittarius
M11	Wild Duck Cluster	open cluster	Scutum
M13	Hercules Cluster	globular cluster	Hercules
M15	—	globular cluster	Pegasus
M22	—	globular cluster	Sagittarius
M27	Dumbbell Nebula	planetary nebula	Vulpecula
M31	Andromeda Galaxy	galaxy	Andromeda
M35	—	open cluster	Gemini
M42	Orion Nebula	gaseous nebula	Orion
M44	Praesepe	open cluster	Cancer
M45	Pleiades	open cluster	Taurus
M57	Ring Nebula	planetary nebula	Lyra
M67	King Cobra Cluster	open cluster	Cancer
IC 2602	Southern Pleiades	open cluster	Carina
NGC 752	—	open cluster	Andromeda
NGC 3242	Ghost of Jupiter	planetary nebula	Hydra
NGC 3372	Eta Carinae Nebula	gaseous nebula	Carina
NGC 4755	Jewel Box	open cluster	Crux
NGC 5139	Omega Centauri	globular cluster	Centaurus

Other objects

In the late eighteenth century Charles Messier, a French astronomer, compiled a catalogue of objects while he was searching for comets. These nebulae, clusters of stars and galaxies were given 'Messier numbers' (some already had names given several thousand years ago, such as Praesepe – the Beehive Cluster). Some, such as the Andromeda galaxy, M31, and the Orion Nebula, M42, may be seen by the naked eye, but all those given in the list will benefit from the use of binoculars. Apart from galaxies, such as M31, which contain thousands of millions of stars, there are also two types of cluster: open clusters, such as M45, the Pleiades, which may consist of a few dozen to some hundreds of stars; and globular clusters, such as M13 in Hercules, which are spherical concentrations of many thousands of stars. One or two gaseous nebulae, consisting of gas illuminated by stars within them, are also visible. The Orion Nebula, M42, is one, and is illuminated by a group of four stars, known as the Trapezium, which may be seen by using a good pair of binoculars.

Our nearest big galaxy is **Andromeda**, or Messier 31 (M31). It lies 2.5 million light years away, meaning a text message (travelling at the speed of light) would take 2.5 million years to reach Andromeda. Andromeda is thought to have 1 trillion stars and it is travelling towards our Milky Way galaxy, expecting to reach us in 4–5 billion years. The Andromeda galaxy is visible to the naked eye on a dark clear night.

Dates and time

Astronomers, worldwide, use a standardized method of expressing the date and time. This prevents confusion when comparing observations made by observers at different longitudes (and time zones). The various elements are given in order of increasing specificity. The date and time are based on the Greenwich meridian (GMT), and ignore any changes for Summer Time/Daylight Saving Time (DST) and any adjustments for local time at the observer's location. This standard is known as Coordinated Universal Time (UTC), generally given as Universal Time (UT). UT is based on the Earth's rotation, which varies over long periods of time, while UTC is based on atomic clocks. Consequently there is a small difference between UT and UTC. All times given in this book are in UT. Experienced astronomers set a (cheap) watch or clock to Universal Time and keep it that way. Smartphone users may use a simple world clock app, provided they lock it to the time on the Greenwich meridian (GMT).

Similarly, the date given for an event is the date as it applies at the Greenwich meridian, i.e., in UT. Occasionally, this may differ from the date as given by your local time. An event that occurs (say) early in the morning in Europe, may seem to occur on the previous day to an observer to the west (such as in the USA), when local time is taken into account. This is another complication that is avoided by using the Universal Time standard.

The mutual gravitational attraction between the Earth and the Moon contributes to irregularities in the Earth's rotation. The **caesium atomic clock** was introduced in 1955. This clock served as a precise timekeeper based on the caesium-133 atom's vibrations. In 1967, the second was officially defined as 9,192,631,770 periods of this vibration and this led to **International Atomic Time (TAI)**, but it was not linked to the Earth's rotation. Coordinated Universal Time (UTC) was therefore established as the basis of international timekeeping. UTC is adjusted occasionally to stay within 0.9 seconds of Universal Time (UT1 or GMT) by the addition of an extra second, thus ensuring UTC's alignment with the Earth's rotation. These adjustments are made in the last minute of December or June, occasionally in March or September. It is the responsibility of the International Earth Rotation and Reference Systems Service (IERS) to apply leap seconds; there have been 27 leap seconds added since the first one was inserted in 1972.

Major Events in 2026

03 Jan.	Earth at perihelion (closest to the Sun)
04 Jan.	Quadrantid meteor shower maximum
07 Jan.	Comet 24P/Schaumasse reaches peak apparent magnitude of around 8
10 Jan.	Jupiter at opposition
08 Feb.	α-Centaurid meteor shower maximum
17 Feb.	Annular solar eclipse (Antarctica), partial eclipse (south Argentina & Chile, south Africa)
19 Feb.	Mercury at greatest eastern elongation
28 Feb.	Mercury, Venus, Jupiter, Saturn, Uranus and Neptune are above the horizon after sunset
03 Mar.	Total lunar eclipse, visible from eastern Asia, Australia, parts of North and South America
08 Mar.	Daylight Saving Time (DST) begins in USA
14 Mar.	γ-Normid meteor shower maximum
20 Mar.	Vernal equinox
29 Mar.	Summer Time begins (BST in UK)
03 Apr.	Mercury at greatest western elongation
05 Apr.	Daylight Saving Time ends (New Zealand & parts of Australia)
22 Apr.	April Lyrid meteor shower maximum
06 May.	η-Aquariid meteor shower maximum
15 Jun.	Mercury at greatest eastern elongation
21 Jun.	Summer solstice
06 Jul.	Earth at aphelion (farthest from the Sun)
28 Jul.	Piscis Austrinid meteor shower maximum
30 Jul.	α-Capricornid meteor shower maximum
30 Jul.	Southern δ-Aquariid meteor shower maximum
02 Aug.	Mercury at greatest western elongation
03 Aug.	Comet 10P/Tempel reaches peak apparent magnitude of around 7
10 Aug.	Mercury, Mars, Jupiter, Saturn, Uranus and Neptune are above the horizon just before sunrise

12 Aug.	Total solar eclipse (Iceland, Spain, Greenland and the Arctic). Partial eclipse visible from North America, west Africa and Europe.
13 Aug.	Perseid meteor shower maximum
15 Aug.	Venus at greatest eastern elongation
28 Aug.	Partial lunar eclipse. Visible from Europe (including London), Africa, the eastern Pacific.
01 Sep.	α-Aurigid meteor shower maximum
23 Sep.	Autumnal equinox
26 Sep.	Neptune at opposition
30 Sep.	Minor planet (192) Nausikaa at opposition
04 Oct.	Saturn at opposition
04 Oct.	Minor planet (2) Pallas at opposition
04 Oct.	Daylight Saving Time begins (parts of Australia and New Zealand)
09 Oct.	Draconid meteor shower maximum
10 Oct.	Southern Taurid meteor shower maximum
12 Oct.	Mercury at greatest eastern elongation
13 Oct.	Minor planet (4) Vesta at opposition
21 Oct.	Orionid meteor shower maximum
25 Oct.	British Summer Time ends (UK reverts to GMT)
01 Nov.	Daylight Saving Time ends in USA
12 Nov.	Northern Taurid meteor shower maximum
18 Nov.	Leonid meteor shower maximum
20 Nov.	Mercury at greatest western elongation
25 Nov.	Uranus at opposition
07 Dec.	Puppid-Velid meteor shower maximum
11 Dec.	Moon at apogee = 406,421 km (most distant of year)
14 Dec.	Geminid meteor shower maximum
21 Dec.	Winter solstice
22 Dec.	Ursid meteor shower maximum
24 Dec.	Moon at perigee = 356,650 km (closest of the year)

The Moon

The Moon at First Quarter.

The Moon

The monthly pages include diagrams showing the *phase* of the Moon (see page 32) for every day of the month, and also indicate the day in the *lunation* (or *age* of the Moon), which begins at New Moon. The diagrams showing the Moon's phase are repeated for southern-hemisphere observers who will see the Moon south up.

The lunar phase cycle has a duration of 29.5 days and starts with the New Moon, when the near side of the Moon facing us is not illuminated by the Sun. The New Moon rises several hours after midnight and sets in the late afternoon or night; the First Quarter Moon rises in the late morning and sets around midnight; the Full Moon is above the horizon most of the night and the Last Quarter Moon rises around midnight and sets in the late morning.

Although the main features of the surface – the light highlands and the dark maria (seas) – may be seen with the naked eye, far more features may be detected with the use of binoculars or any telescope. The many craters are best seen when they are close to the *terminator* (the boundary between the illuminated and the non-illuminated areas of the surface), when the Sun rises or sets over any particular region of the Moon and the crater walls or central peaks cast strong shadows. These are best seen at crescent, quarter and gibbous phases. Most features become difficult to see at Full Moon, although this is the best time to see the bright ray systems surrounding certain craters. Accompanying the Moon map on the following pages is a list of prominent features, including the days in the lunation when they are normally close to the terminator and thus easiest to see.

The dates of visibility vary slightly through the effects of *libration*. Because the Moon's orbit is inclined to the Earth's equator and also because it moves in an ellipse, the Moon appears to rock slightly from side to side (and nod up and down). Over time a total of 59 per cent of the lunar terrain can be observed. Areas near the *limb* (the edge of the Moon) may vary considerably in their location and visibility. This is easily noticeable with Mare Crisium (Sea of Crises, eastern limb of the Moon, in the western sky) and the craters Tycho (southern part of the Moon) and Plato (northern part). Another effect is that at crescent phases before and after New Moon, the normally non-illuminated portion of the Moon receives a certain amount of light, reflected from the Earth.

This *Earthshine* may enable certain bright features (such as the craters Aristarchus, Kepler and Copernicus on the western limb of the Moon, in the eastern sky) to be detected.

Moon features

Abulfeda	6:20	Gassendi	11:25	Philolaus	9:23
Agrippa	7:21	Geminus	3:17	Piccolomini	5:19
Albategnius	7:21	Goclenius	4:18	Pitatus	8:22
Aliacensis	7:21	Grimaldi	13–14:27–28	Pitiscus	5:19
Alphonsus	8:22	Gutenberg	5:19	Plato	8:22
Anaxagoras	9:23	Hercules	5:19	Plinius	6:20
Anaximenes	11:25	Herodotus	11:25	Posidonius	5:19
Archimedes	8:22	Hipparchus	7:21	Proclus	14:18
Aristarchus	11:25	Hommel	5:19	Ptolemaeus	8:22
Aristillus	7:21	Humboldt	3:15	Purbach	8:22
Aristoteles	6:20	Janssen	4:18	Pythagoras	12:26
Arzachel	8:22	Julius Caesar	6:20	Rabbi Levi	6:20
Atlas	4:18	Kepler	10:24	Reinhold	9:23
Autolycus	7:21	Landsberg	10:24	Rima Ariadaeus	6:20
Barrow	7:21	Langrenus	3:17	Rupes Recta	8
Billy	12:26	Letronne	11:25	Saussure	8:22
Birt	8:22	Linné	6	Scheiner	10:24
Blancanus	9:23	Longomontanus	9:23	Schickard	12:26
Bullialdus	9:23	Macrobius	4:18	Sinus Iridum	10:24
Bürg	5:19	Mädler	5:19	Snellius	3:17
Campanus	10:24	Maginus	8:22	Stöfler	7:21
Cassini	7:21	Manilius	7:21	Taruntius	4:18
Catharina	6:20	Mare Crisium	2–3:16–17	Thebit	8:22
Clavius	9:23	Maurolycus	6:20	Theophilus	5:19
Cleomedes	3:17	Mercator	10:24	Timocharis	8:22
Copernicus	9:23	Metius	4:18	Triesnecker	6–7:21
Cyrillus	6:20	Meton	6:20	Tycho	8:22
Delambre	6:20	Mons Pico	8:22	Vallis Alpes	7:21
Deslandres	8:22	Mons Piton	8:22	Vallis Schröteri	11:25
Endymion	3:17	Mons Rümker	12:26	Vlacq	5:19
Eratosthenes	8:22	Montes Alpes	6–8:21	Walther	7:21
Eudoxus	6:20	Montes Apenninus	8	Wargentin	12:27
Fra Mauro	9:23	Orontius	8:22	Werner	7:21
Fracastorius	5:19	Pallas	8:22	Wilhelm	9:23
Franklin	4:18	Petavius	3:17	Zagut	6:20

Map of the Moon

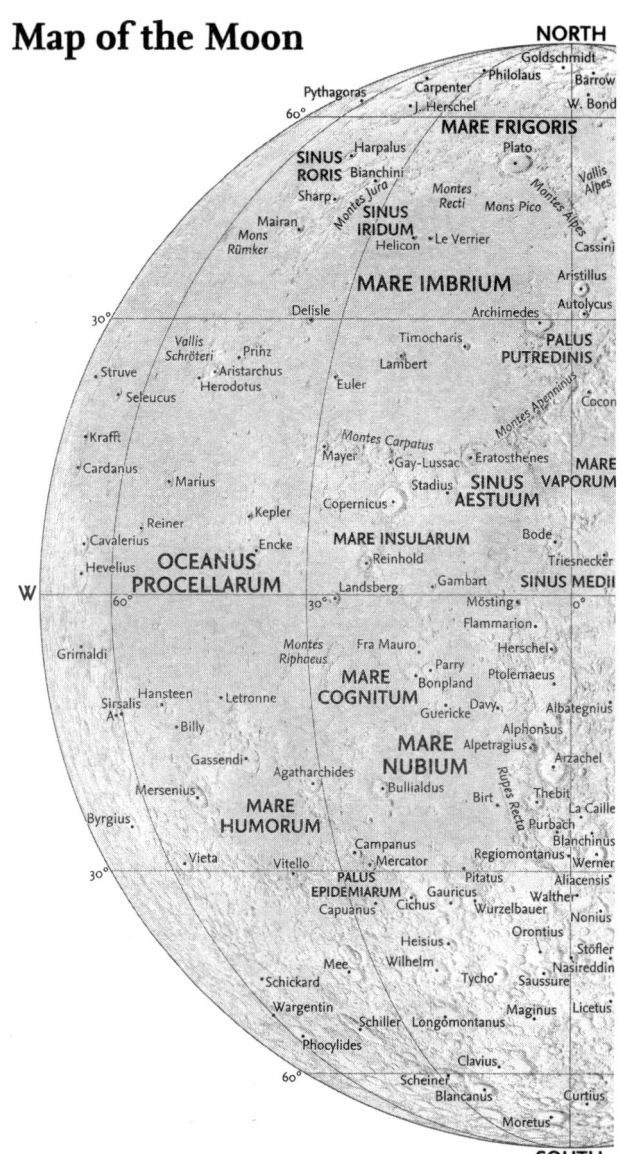

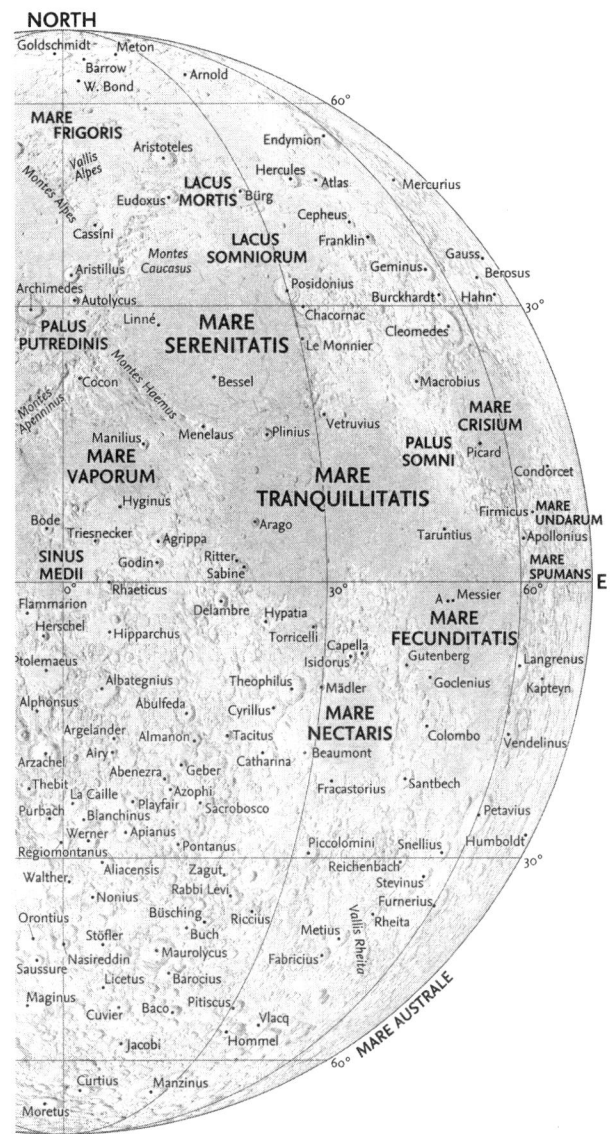

NORTH

Goldschmidt · Meton
Barrow · Arnold
W. Bond
60°
MARE
FRIGORIS
Endymion
Aristoteles
Vallis Alpes
Montes Alpes
Hercules · Atlas · Mercurius
LACUS
Eudoxus· MORTIS Bürg
Cepheus
Cassini
LACUS Franklin
Montes SOMNIORUM Gauss
Aristillus Caucasus Geminus· Berosus
Archimedes Posidonius Burckhardt Hahn·
·Autolycus Chacornac 30°
PALUS Linné· MARE Cleomedes
PUTREDINIS SERENITATIS Le Monnier
·Cocon ·Bessel ·Macrobius
Menelaus MARE
Montes Haemus CRISIUM
Manilius, Menelaus ·Plinius Vetruvius PALUS Picard
MARE Hyginus SOMNI Condorcet
VAPORUM MARE MARE
Bode TRANQUILLITATIS UNDARUM
Triesnecker ·Arago Firmicus· Apollonius
SINUS Godin· Taruntius MARE
MEDII Ritter· SPUMANS E
Sabine
Rhaeticus 30° 60°
Flammarion Delambre Hypatia A.·Messier
Herschel ·Hipparchus Torricelli MARE
Ptolemaeus Isidorus FECUNDITATIS
Capella Gutenberg Langrenus
·Albategnius Theophilus ·Mädler Goclenius Kapteyn
Alphonsus Abulfeda Cyrillus· MARE
·Argelander Almanon· ·Tacitus NECTARIS Colombo Vendelinus
Airy· Geber ·Beaumont
Arzachel Abenezra Catharina
·Thebit La Caille ·Azophi Fracastorius ·Santbech
Purbach Blanchinus Sacrobosco Petavius
Werner ·Apianus Piccolomini Snellius Humboldt
Regiomontanus ·Pontanus 30°
Walther, ·Aliacensis Zagut, Reichenbach
·Nonius Rabbi Levi Stevinus·
Orontius Büsching Riccius Furnerius
Stöfler ·Buch Metius ·Rheita
Saussure Nasireddin ·Maurolycus Fabricius·
Maginus Licetus ·Barocius
Cuvier Baco· Pitiscus,
·Jacobi Vlacq
Curtius Hommel
Moretus Manzinus
60° MARE AUSTRALE

SOUTH

Moon phases

The diagram on this page shows how the apparent shape of the Moon changes over a cycle that lasts 29.5 days. The phases are purely related to the position of the Moon in its orbit around the Earth. They are not, as some people mistakenly believe, caused by the shadow of the Earth on the Moon. The only time that happens is during a rare lunar eclipse, as described on pages 10–11. The diagram shows that New Moon is when the Moon lies between the Sun and the Earth, and Full Moon when the Earth is between the Sun and the Moon. The *age* of the Moon (the day in the *lunation*) begins at New Moon, which may be used to determine the best time for observation of specific lunar features.

After New Moon we have a waxing crescent until First Quarter, after which the Moon is described as waxing gibbous. After Full Moon we have a waning gibbous Moon until Last Quarter. Following that, we have a waning crescent until the next New Moon, when the sequence repeats. The dates when the Moon is closest to the Earth (at *perigee*) and farthest from it (at *apogee*) are shown in the monthly calendars. The changing distance is due to the Moon's orbit being non-circular.

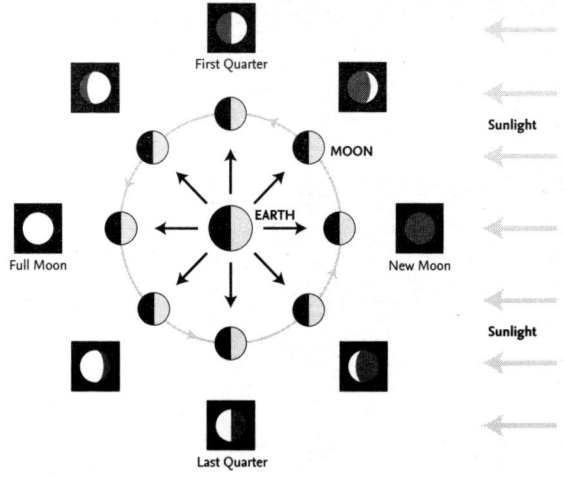

The Circumpolar Constellations

The northern circumpolar constellations
The circumpolar stars are the key to starting to identify the
constellations. The charts on pages 34 and 37 show the northern
and southern circumpolar constellations respectively. For
anyone in the northern hemisphere the northern circumpolar
constellations are visible at any time of the year, and nearly
everyone is familiar with the seven stars of the *Plough* – known
as the *Big Dipper* in North America – an asterism that forms
part of the large constellation of *Ursa Major* (the Great Bear).

Ursa Major
Due to the Earth's orbit around the Sun, Ursa Major lies in
different parts of the evening sky at different periods of the
year. The diagram below shows its position at the beginning
of the four main seasons. The seven stars of the Plough or
Big Dipper remain visible throughout the year anywhere north
of latitude 40°N.

*The position of
the Big Dipper
(the Plough)
throughout the
year in relation
to the northern
horizon and
Polaris, the
Pole Star.*

The northern circumpolar constellations.

Polaris and Ursa Minor

The two stars **Dubhe** and **Merak** (α and β Ursae Majoris, respectively), farthest from the 'tail' are known as the Pointers. A line from Merak to Dubhe, extended about five times their separation, leads to the Pole Star, **Polaris**, or α Ursae Minoris. All the stars in the northern sky appear to rotate around it.

There are five main stars in the constellation of **Ursa Minor**, and the two farthest from the Pole, **Kochab** and **Pherkad** (β and γ Ursae Minoris, respectively), are known as the Guards.

Cassiopeia
Facing Ursa Major on the other side of Polaris lies **Cassiopeia**. It is highly distinctive, appearing as five stars forming a letter 'W' or 'M' depending on its orientation. Provided the sky is reasonably clear of clouds, you will nearly always be able to see either Ursa Major or Cassiopeia, and thus be able to orientate yourself on the sky. To find Cassiopeia, start with **Alioth** (ε Ursae Majoris), the first star in the tail of the Great Bear. A line from this star extended through Polaris points directly towards γ Cassiopeiae, the central star of the five.

Cepheus
Although the constellation of **Cepheus** is fully circumpolar, it is not nearly as well-known as Ursa Major, Ursa Minor or Cassiopeia, partly because its stars are fainter. Its shape is rather like the gable end of a house. The line from the Pointers through Polaris, if extended, leads to **Errai** (γ Cephei) at the 'top' of the 'gable'. The brightest star, **Alderamin** (α Cephei) lies in the Milky Way region, at the 'bottom right-hand corner' of the figure.

Draco
The constellation of **Draco** consists of a quadrilateral of stars, known as the Head of Draco (and also the Lozenge), and a long chain of stars forming the neck and body of the dragon. To find the Head of Draco, locate the two stars **Phecda** and **Megrez** (γ and δ Ursae Majoris) in the Plough, opposite the Pointers. On the opposite side of the sky to the head of Draco, the whole of the faint constellation of **Camelopardalis** is visible.

For observers slightly farther north, say at 50°N, additional constellations become circumpolar. The most important of these are **Perseus** (not far from Cassiopeia and mostly visible) and, farther round, the northern portion of **Auriga**, with bright *Capella* (α Aurigae). On the other side of the sky is *Deneb*, the brightest star in **Cygnus**, although it is often close to the horizon, especially during the early night in the winter months. *Vega* (α Lyrae), another of the three stars that form the Summer Triangle, is even farther south, often brushing the northern horizon, and only truly circumpolar and clearly seen at any time of the year for observers at 60°N.

Such far northern observers will also find that **Castor** (α Geminorum) is actually circumpolar, although at times it is extremely low on the horizon. The other bright star in **Gemini**, *Pollux* (β Geminorum) is slightly farther south and cannot really be considered circumpolar.

The spin of the Earth results in the 24-hour cycle of successive sunrises and sunsets, moonrises and moonsets and the sky moving east to west. Long-exposure photography captures this motion, where an exposure time of 15 minutes or more produces images of star trails; the darker the sky the longer the exposure (away from light pollution) and the longer the trails. These tremendous arcs of light are a reminder of the continuous motion of the planet.

The southern circumpolar constellations

For anyone in the southern hemisphere the striking pattern of four stars that make up the constellation of **Crux** (the Southern Cross) is the key constellation to look for, although at times it may be brushing the horizon for observers at 30°S – roughly the latitude of Sydney in Australia. Nearby are the two stars **Rigil Kentaurus** and *Hadar* (α and β Centauri, respectively). Both stars are actually triple star systems. One of the three stars of Rigil Kentaurus is called Proxima Centaurus; it is the closest star to Earth, lying 4.2 light-years away. This star has an Earth-sized planet which orbits the star every 11.2 days. These circumpolar stars are visible throughout the year for most observers, although for observers farther north, the stars may become difficult to see as they are low on the horizon in the months of October and November.

The southern circumpolar constellations.

Crux

The distinctive shape of the constellation of **Crux** is usually easy to identify, although some people (especially northerners unused to the southern sky) may wrongly identify the slightly larger False Cross, formed by the stars **Aspidiske** and **Avior** (ι and ε Carinae respectively) plus **Markeb** and **Alsephina** (κ and δ Velorum). The Coalsack Nebula (a vast dark cloud of obscuring dust up to 35

The position of Crux, the Southern Cross, throughout the year in relation to the southern horizon. The two brightest stars in Centaurus are also shown.

light-years across) is readily visible on the southern side of Crux between **Acrux** and **Mimosa** (α and β Crucis).

A line through **Gacrux** (γ Crucis) and **Acrux** (α Crucis) points approximately in the direction of the South Celestial Pole which lies in the faint constellation of **Octans**, crossing the faint constellations of **Musca** and **Chamaeleon**. The basic triangular shape of Octans itself is best found by extending a line from **Peacock** (α Pavonis) in the constellation **Pavo**, through β Pavonis, by about the same distance as that between the stars.

Centaurus

Although Crux is a distinctive shape, **Centaurus** is a large, rather straggling constellation, with one notable object: the giant, bright globular cluster **Omega Centauri** (the brightest and largest globular in the sky consisting of around 10 million stars), which lies towards the north, on the line from **Hadar** (β Centauri) through ε Centauri. A more accurate method of locating the South Celestial Pole is to use the line from Crux and imagine a line at right angles and midway to the line between Hadar and Rigil Kentaurus – the intersection of these two lines identifies the SCP.

Carina

Apart from the two stars that form part of the False Cross, the constellation of Carina is, like Centaurus, a large, sprawling constellation. It contains one striking open cluster, the **Southern Pleiades**, and a remarkable emission nebula, the **Eta Carinae Nebula**. The second brightest star in the sky (after Sirius) is **Canopus**, α Carinae, which lies far away to the west.

The Magellanic Clouds

On the opposite side of the South Celestial Pole to Crux and Centaurus lie the two Magellanic Clouds: small irregular satellite galaxies of our Milky Way. The **Small Magellanic Cloud** (SMC) lies a distance of 200,000 light-years from Earth, it sits to one side of the triangular constellation of **Hydrus**, but it is actually within the constellation of **Tucana**.

Nearby is another bright globular cluster, **47 Tucanae**. Hydrus itself is also easily identified from the star **Achernar**, α Eridani, the rather isolated brilliant star at the southern end of **Eridanus**, which wanders a long way south, having begun at the foot of Orion.

The **Large Magellanic Cloud** (LMC) is around 40,000 light-years closer to Earth than the SMC and it lies within the faint constellation of **Dorado**. The LMC contains the large, readily visible **Tarantula Nebula**, an emission nebula that is a major star-forming region. The LMC has an apparent size of around 10° in the sky, equivalent to 20 Moons across.

The **Large and Small Magellanic Clouds** are satellite galaxies that never set in the southern hemisphere due to their proximity to the South Celestial Pole. Named *al-Bakr* ('the white ox') by Persian astronomer Abd al-Rahman al-Sufi in 964 CE, the Large Magellanic Cloud – closer to the Earth at 160,000 light-years than its counterpart, the Small Magellanic Cloud – contains about 30 billion stars. The Small Magellanic Cloud has one tenth of the number of stars of the LMC. These galaxies formed around the same time as the Milky Way. Recent studies, like the **Dark Energy Survey**, suggest they may eventually merge due to their interactions.

The Monthly Maps

The charts in this book are designed to be used more or less anywhere in the world. They are not suitable for use at very high northern or southern latitudes (beyond 60°N or 60°S). That is slightly less than the latitudes of the Arctic and Antarctic Circles, beyond which there are approximately six months of daylight, followed by six months of darkness. The design may seem a little complicated, but these diagrams should make their usage clear. The main charts are given in pairs, one pair for each month: Looking North and Looking South.

Obviously, the region of the sky that is visible at any time entirely depends on one's location on Earth. If you look horizontally towards the horizon and then look vertically up at the zenith, your line of sight will have moved through an angle of 90°. Now imagine flattening this 90° field of view into a rectangular 'window'; this corresponds to a particular portion of the charts. To find the portion of the sky that is visible from your latitude, you just need to isolate the relevant part of the chart.

Lines of latitude are defined in the right-hand and left-hand margins. The base of your viewing 'window' should lie horizontally along the line that represents your latitude; anything below this on the chart will be out of view below the horizon. The top edge indicates your zenith; anything above this on the chart will be behind your head. The diagram on the next page shows this 'window' highlighted for the latitude of 50°N, facing the northern horizon.

The two diagrams on page 42 show just the viewing 'windows' for the northward view (top diagram) and southward view (bottom diagram) at a latitude of 40°N (the latitude of Philadelphia in the United States or Madrid in Spain); the second pair on page 43 shows the same two views from latitude 30°S (the latitude of Durban in South Africa). The charts show the sky at the following times: 23:00 at the start of the month, 22:00 in the middle of the month and 21:00 at the end of the month.

To help you choose the correct latitude, there is a world map on pages 44–45.

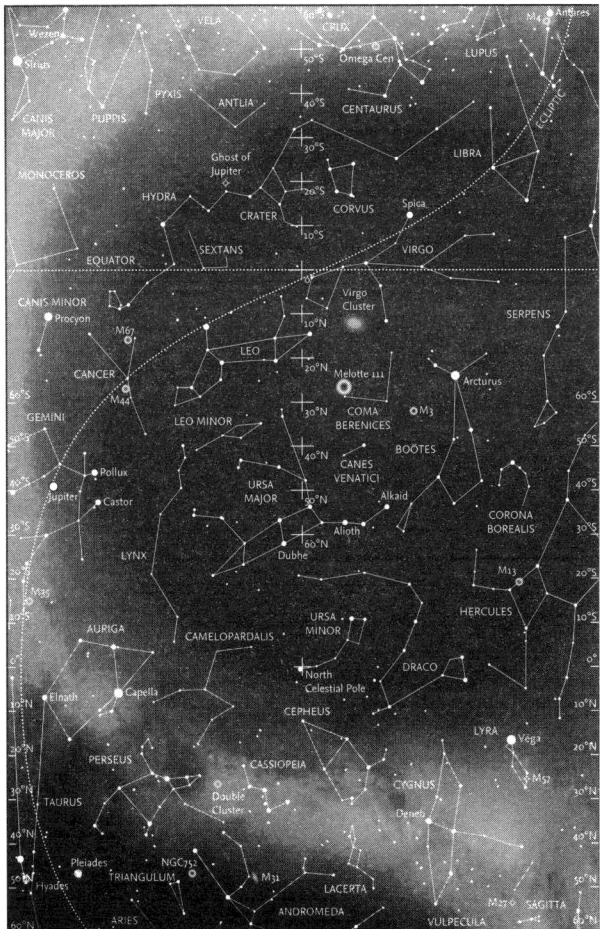

Horizon window, from the northern horizon (solid line at the bottom) to the zenith (the dotted line) for the latitude of 50°N.

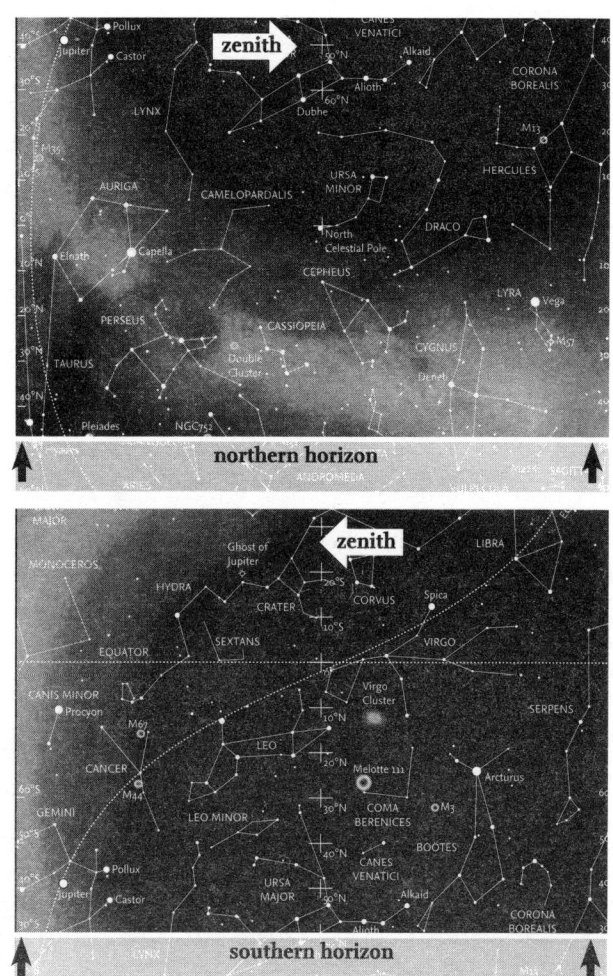

Horizons for latitude 40°N.

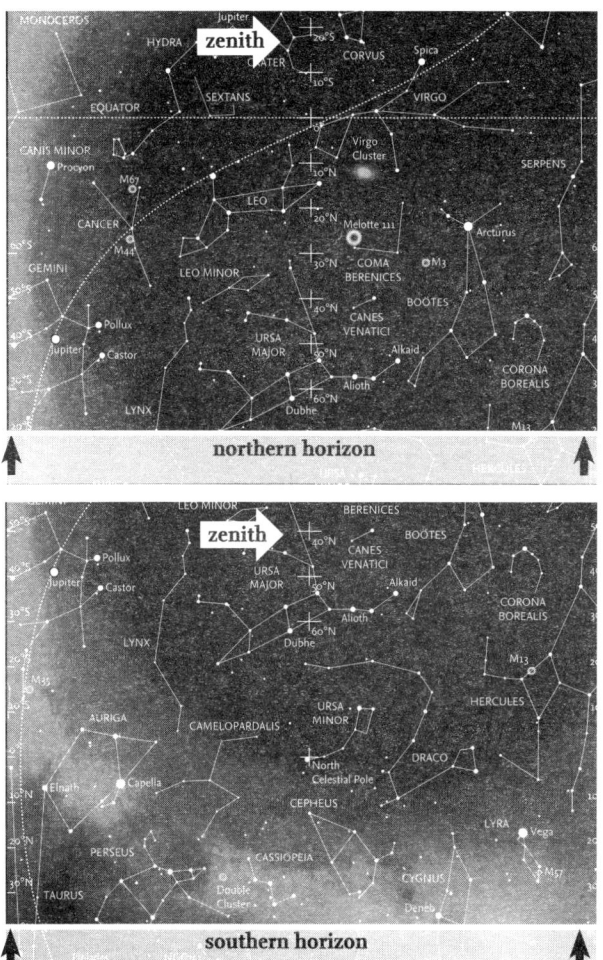

Horizons for latitude 30°S.

World Map

Murmansk

70°N

ockholm
lon
Amsterdam
Moscow
60°N

Berlin

ASIA
50°N

EUROPE

Paris

Vladivostok

Rome
Beijing
40°N

Tokyo

Teheran
Shanghai
30°N

Cairo
Delhi

Hong Kong
20°N

Mumbai

Khartoum
Manila

Bangkok
Guam
10°N

agos
Addis Ababa
Marshall
Islands

AFRICA
Singapore

Equator
0°

nshasa
Nairobi
New Guinea

Dar es Salaam
Jakarta
10°S

OCEANIA

Darwin
20°S

Madagascar

Australia
30°S

Perth
Sydney

Cape Town
Wellington

New
Zealand
50°S

Tips for Observing the Night Sky

A successful observing session requires planning, flexibility and patience. Check the weather forecast, sunset and moonrise and moonset times and choose a location where you have a clear view of the horizon wherever possible. Bring your star guide or app on a mobile device, a red torch to aid your night vision and a pair of binoculars and a tripod to keep them steady. You may be outside for a while – it's advised that you wear warm clothes, socks and shoes and bring hot drinks if it's a cold night.

If you are planning on looking for meteors or looking up at the night sky for a while you may wish to bring a deckchair or picnic blanket with you. When looking at faint objects (unlike the Moon) you will need to give your eyes 10–20 minutes to adapt to the dark and achieve night vision. If you are using your phone switch it to night vision mode and use a red torch to find your way around so that you maintain your night vision.

Astrophotography

You can capture the Moon and five planets visible to the naked eye or the brightest stars using your smartphone. The challenge is keeping your phone steady and gaining control over the focus and exposure (if the camera allows it). It is advised that you attach the phone to a tripod, or you could use an adaptor to attach your phone to the eyepiece of a telescope – this will allow you to align it correctly with minimal effort. You can use the zoom function on your camera to change the magnification of the image.

There are many apps available that will provide more control of your camera settings and allow you to take better quality images: NightCap Pro, ProCam and AstroShader for iOS; and Open Camera, DeepSkyCamera or Camera FV-5 for Android. For the Moon and bright planets your camera should auto lock and focus, otherwise manually set the focus to infinity and keep the exposure short (a fraction of a second if it is a gibbous or Full Moon, one or two seconds for a thin crescent Moon and planets). You could control the shutter remotely via the volume control on wireless headphones, or set a timer on your phone (3 seconds delay). For fainter objects increase the exposure and the ISO (the sensitivity of the camera detector); however, a higher ISO setting will introduce more noise. Trial and error will help you choose the best settings. Some apps allow images to be stacked, meaning you can overlay images to increase brightness further, useful for the constellations, clusters such as the Pleiades and nebulous objects such as the Orion Nebula and Andromeda.

January

January – Introduction

The Earth

The Earth reaches perihelion (the closest point to the Sun in its yearly orbit) on 3 January at 17:15 Universal Time, when it is at a distance of 0.9833 AU (147,099,900 km). Due to the Earth's elliptical orbit, its distance from the Sun varies by around 3 per cent throughout the year.

Meteors

The **Quadrantid** meteor shower begins on 28 December 2025 and continues until 12 January 2026. It is one of the strongest and most consistent meteor showers. It comes to maximum on the night of 3 and 4 January, during a Full Moon, so moonlight will reduce visibility. The meteors are bright, bluish- or yellowish-white and may reach a maximum rate of 120 per hour. The parent object is minor planet 2003 EH_1.

The shower is named after the former constellation **Quadrans Muralis** (the Mural Quadrant), an early form of astronomical instrument. The Quadrantid radiant, marked on the chart, is now within the northernmost part of **Boötes**, roughly halfway between θ Boötis and τ Herculis.

Comets

Although comets may occasionally become very striking objects in the sky, their occurrence and particularly the existence or length of any tail and their overall magnitude are notoriously difficult to predict. Naturally, it is only possible to predict the return of periodic comets (whose names have the prefix 'P'). Many comets appear unexpectedly (these have names with the prefix 'C'). Most periodic comets are faint and only a very small number ever become bright enough to be easily visible with the naked eye or with binoculars.

Comet 24P/Schaumasse will be visible most of the night from January to March, brightening from magnitude 9 to 8, moving through the constellations of Virgo, Boötes and Serpens Caput. It will make its closest approach on 4 January 2026, passing within a distance equivalent to 60% of the Earth–Sun distance. It is visible at dawn from both hemispheres.

The planets

Mercury starts the month visible just before sunrise and trails the Sun in the evening after 21 January, brightening to mag. –1.3. *Venus* leads the Sun at dawn at the start of the year and transitions to the early evening sky after 6 January, visible at the end of the month at mag. –3.9. *Mars* approaches the Sun in Sagittarius and swings into the dawn sky; however, it is too close to the Sun to be seen (mag. 1.1 to 1.0, then dimming to 1.2). *Jupiter* lies in *Gemini*, it is best seen at opposition at midnight on 10 January, mag. –2.7. *Saturn* moves from *Aquarius* to *Pisces* during the month (mag. 1.0), *Neptune* is nearby. *Uranus* sits in *Taurus* until a few hours after midnight, at mag. 5.6. *Neptune* is in *Pisces* at mag. 7.8.

The origin of moonlight: Zhang Heng

The Chinese astronomer Zhang Heng (78–139 CE) proposed that the Moon's surface reflected rays from the Sun 'like water'. He stated that, 'The Sun is like fire and the Moon like water. The fire gives out light and the water reflects it.' He added that, 'the side that faces the Sun is fully lit, and the side that is away from it is dark.' He realized that the planets also reflected sunlight and could obstruct sunlight as seen from Earth and he explained the role of the Moon and Earth in creating the phenomena of lunar and solar eclipses.

The first spacecraft to land on the far (dark) side of the Moon was the Chinese lander *Chang'e-4*, which landed on 3 January 2019 near the lunar south pole. In December 2020, the *Chang'e-5* mission brought back samples of lunar rock and soil from a volcanic site on the bright near side.

The Chang'e-4 *lander imaged by the* Yutu-2 *rover on the far (dark) side of the Moon, January 2019.*

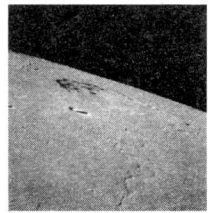

Mons Rümker in Oceanus Procellarum (the Ocean of Storms) – the site of the Chang'e-5 *rover.*

Sunrise and sunset

City	Date	Sunrise	Sunset
Buenos Aires, Argentina			
	Jan. 01	08:45	23:10
	Jan. 31	09:13	23:01
Cape Town, South Africa			
	Jan. 01	03:39	18:01
	Jan. 31	04:07	17:52
London, UK			
	Jan. 01	08:06	16:02
	Jan. 31	07:40	16:48
Los Angeles, USA			
	Jan. 01	14:59	00:55
	Jan. 31	14:51	01:23
Nairobi, Kenya			
	Jan. 01	03:30	15:42
	Jan. 31	03:41	15:51
Sydney, Australia			
	Jan. 01	18:47	09:09
	Jan. 31	19:15	09:01
Tokyo, Japan			
	Jan. 01	21:51	07:38
	Jan. 31	21:42	08:07
Washington, DC, USA			
	Jan. 01	12:27	21:57
	Jan. 31	12:15	22:29
Wellington, New Zealand			
	Jan. 01	16:51	07:57
	Jan. 31	17:25	07:43

NB: the times given are in Universal Time (UT)

The Moon's phases and ages

Northern Hemisphere

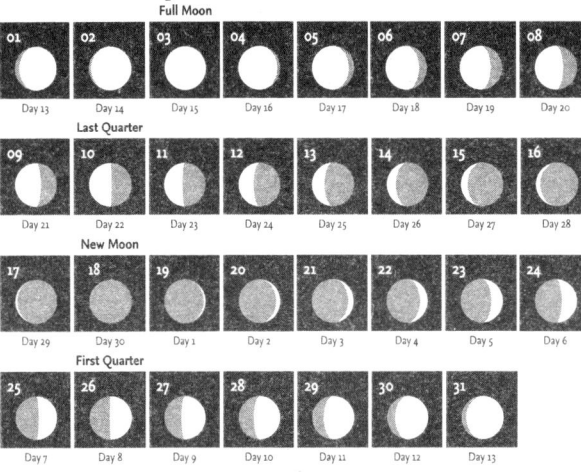

Southern Hemisphere

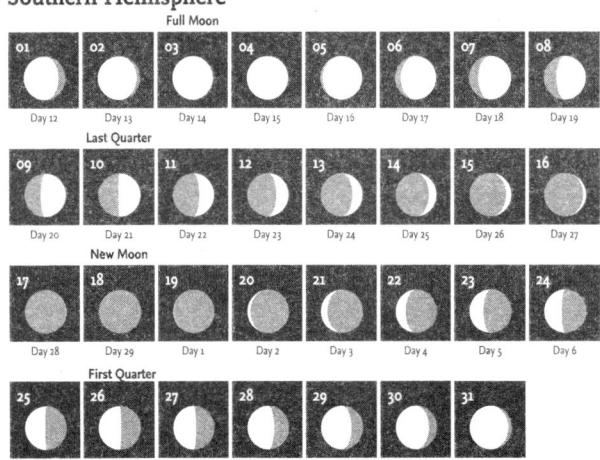

The Moon

The Moon in January

The Moon is Full on 3 January, *Jupiter* will appear 3.7°S of the Moon at mag. –2.7, both are visible from the early evening. A day later *Pollux* will be 3.0°N of the waning gibbous Moon. On 6 January *Regulus* lies only 0.5°S of the Moon. On 10 January *Spica* lies 1.6°N of the Last Quarter Moon, four days later the red-orange star *Antares* will be placed 0.6°N of the waning crescent Moon, appearing low in the east a few hours before sunrise. The lunar cycle resets on 18 January with a New Moon. On 23 January *Saturn* (mag. 1.0) will appear 4.3°S of the waxing crescent Moon in the early evening, both setting shortly after 20:30. Three days later the Moon will be First Quarter; a day later the *Pleiades* cluster lies 1.1°S of the Moon. The month ends with *Jupiter* (mag. –2.6) 3.8°S of the waxing gibbous Moon and *Pollux* lies 3.0°N.

Cygnus X-1: The black hole and the blue supergiant

Until the first direct image of a black hole was published in April 2019, the existence of black holes was determined indirectly, by their influence on their surrounding environment. In 1964, a strong source of X-rays was discovered in the constellation Cygnus. In January 1972, results confirming the presence of a black hole called **Cygnus X-1** locked in a binary system with a blue supergiant star called **HDE 226868** were published by Louise Webster, Paul Murdin (both at the Royal Greenwich Observatory) and Tom Bolton (at the University of Toronto's David Dunlap Observatory).

The blue supergiant star has a surface temperature of around 31,000°C. In contrast, our Sun has a temperature of 5,500°C. Its mass is 20–40 solar masses and it is at least 300,000 times brighter than the Sun. Blue supergiants have considerably shorter lifetimes than low-mass stars. This star is shedding a Sun's worth of gas every 400,000 years, and will use up the fuel in its core at a much faster rate than the Sun, eventually exploding as a supernova, and the remnant will collapse, either into a neutron star or a black hole.

The astronomers analysed the spectrum of starlight, which exhibited periodic shifts due to the orbit of the star around

Cygnus X-1. Their measurements of the orbit of the star revealed the companion to be too massive to be a neutron star, only a black hole could be responsible; recent measurements yield a mass of 21 solar masses. The two objects orbit their centre of mass every 5.6 days; this compact binary system lies 7,300 light-years away.

The black hole pulls matter from the star into a tight swirling accretion disc. The rotating charged matter radiates X-rays that can be detected by telescopes above the Earth's atmosphere. Any particles or light rays that pass within 44 km of the centre are pulled in by the powerful gravitational field beyond the point of return; to escape the black hole they would need to exceed the speed of light, which is impossible.

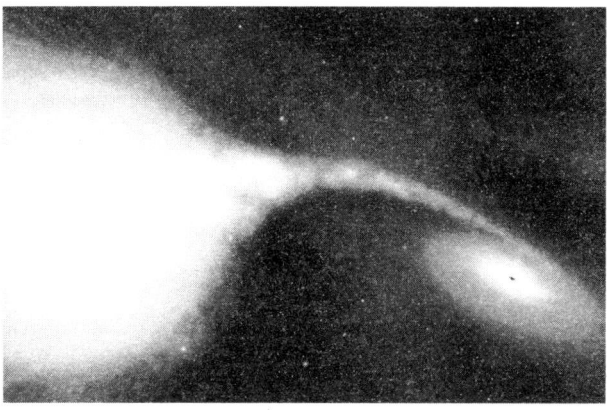

An artist's impression of the binary system of the blue supergiant HDE 226868 and the black hole Cygnus X-1.

Calendar for January

01	21:43	Moon at perigee = 360,348 km
03	10:03	Full Moon
03	17:15	Earth at perihelion (147,099,900 km = 0.9833 AU)
03–04		Quadrantid meteor shower maximum
03	22:01	Jupiter (mag. –2.7) 3.7°S of the Moon
04	03:28	Pollux 3.0°N of the Moon
06	16:20	Regulus 0.5°S of the Moon. An occultation will be visible from Mexico, Australia and New Zealand.
10	08:34	Jupiter (mag. –2.7) at opposition
10	15:48	Last Quarter Moon
10	23:50	Spica 1.6°N of the Moon
13	20:48	Moon at apogee = 405,437 km
14	19:28	Antares 0.6°N of the Moon. An occultation will be visible from Australia.
18	19:52	New Moon
23	12:31	Saturn (mag. 1.0) 4.3°S of the Moon
26	04:47	First Quarter Moon
27	21:07	Pleiades 1.1°S of the Moon
29	21:53	Moon at perigee = 365,878 km
31	02:31	Jupiter (mag. –2.6) 3.8°S of the Moon
31	13:45	Pollux 3.0°N of the Moon

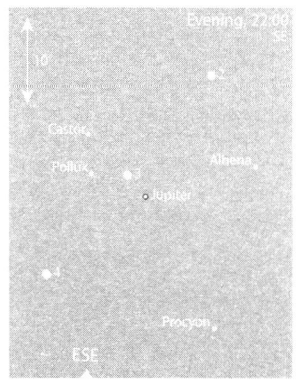

3–4 January: *The Full Moon lies in Gemini close to Jupiter (mag. –2.7) and Pollux. Castor can be seen above the Moon (as seen from London).*

15 January: *Red-orange Antares sits next to the waning crescent Moon in the dawn sky, low in the southeast (as seen from Sydney).*

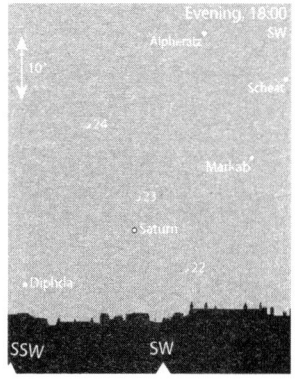

23 January: *The waxing crescent Moon lies above Saturn. The stars of Diphda (β Cet) and Markab (α Peg) are close by (as seen from London).*

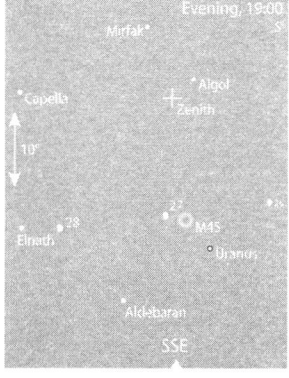

27 January: *The waxing gibbous Moon is next to the Pleiades (M45) and Uranus. Aldebaran, Capella and Elnath (β Cet) appear in the same part of the sky (as seen from central USA).*

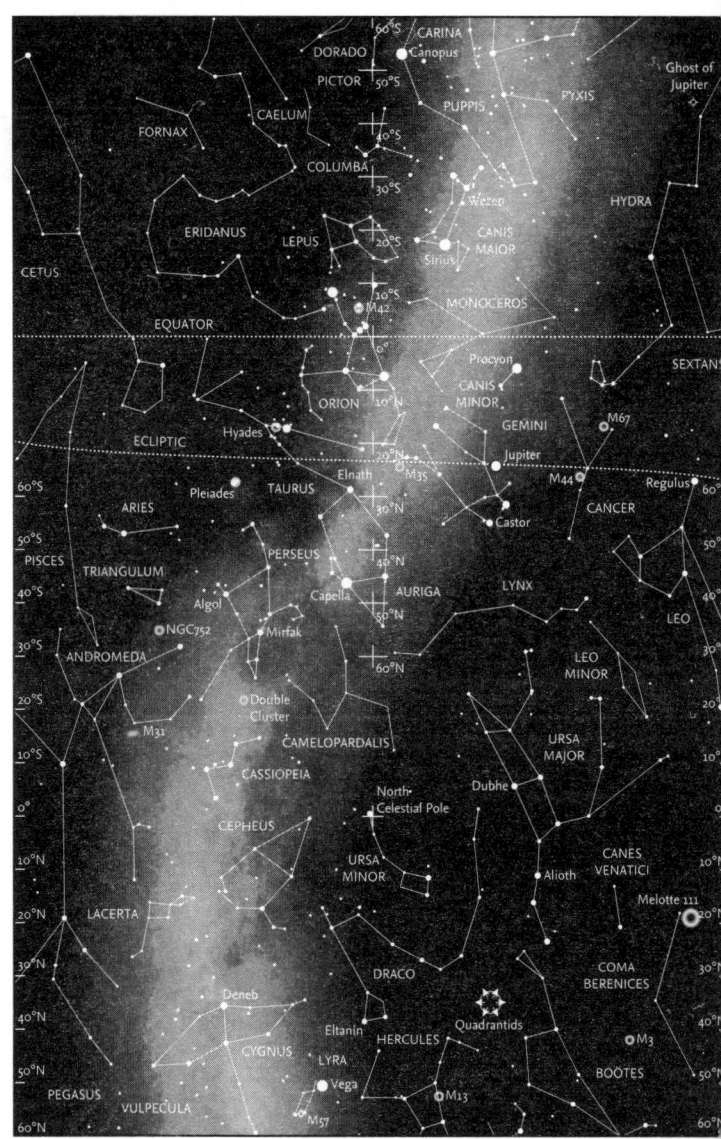

January – Looking North

Most of the important circumpolar constellations are easy to see in the northern sky at this time of year. **Ursa Major** stands more-or-less vertically above the horizon in the northeast, with the zodiacal constellation of **Leo** rising in the east. To the north, the stars of **Ursa Minor** lie below **Polaris** (the Pole Star). The head of **Draco** and its brightest star **Eltanin** (γ Draconis) are low on the northern horizon and visible to those at high latitudes (above 40°N); the stars may be difficult to see unless observing conditions are good. The blue supergiant **Deneb** (α Cygni) or the even brighter **Vega** (α Lyrae) are visible farther south. Both **Cepheus** and **Cassiopeia** are readily visible in the northwest, and even the faint constellation of **Camelopardalis** is high enough in the sky for it to be easily visible.

For observers between about 30 and 50°N, near the zenith is the constellation of **Auriga** (the Charioteer), with brilliant **Capella** (α Aurigae), directly overhead. Slightly to the west of Capella lies a small triangle of fainter stars, known as the Kids. (Ancient mythological representations of Auriga show him carrying two young goats.) Together with the northernmost bright star in **Taurus, Elnath** (β Tauri), the body of Auriga forms a large pentagon on the sky, with the Kids lying on the western side. Farther down towards the west are the constellations of **Perseus,** with the variable star **Algol**, and **Andromeda**. The Great Square of **Pegasus** is approaching the horizon.

On 19 January 2024, the Japan Aerospace Exploration Agency (JAXA) successfully landed their Moon Sniper – the **Smart Lander for Investigating Moon (SLIM)** – near Shioli crater at the Sea of Nectar (Mare Nectaris). The lander used facial recognition technology to identify lunar craters as it approached the lunar surface, and it landed within 100 metres of its target point.

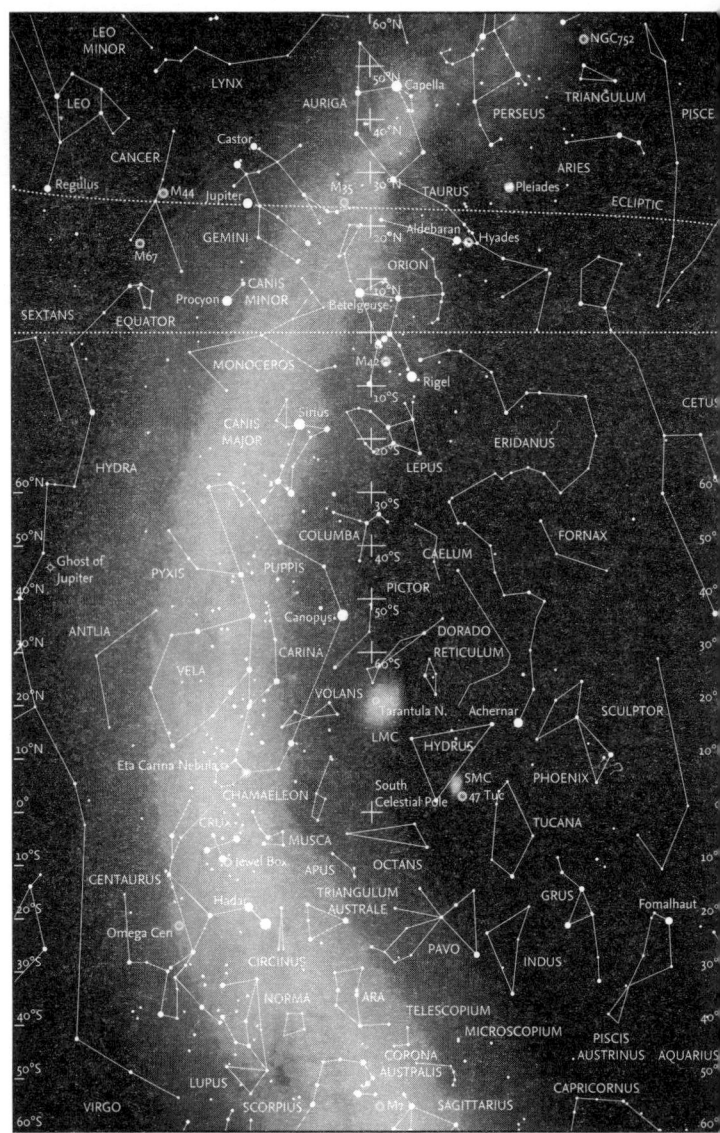

January – Looking South

The southern sky is dominated by **Orion**, prominent during the winter months, visible at some time during the night. It is highly distinctive, with a line of three stars that form Orion's Belt. To most observers, the bright star **Betelgeuse** (α Orionis) shows a reddish tinge; it is known as a red supergiant star, in contrast to the brilliant bluish-white **Rigel** (β Orionis). The three stars of the belt lie directly south of the celestial equator. A vertical line of three 'stars' forms the 'sword' that hangs south of the 'belt'. With good viewing, the central 'star' appears as a hazy spot, even to the naked eye, and is actually the **Orion Nebula**, a star formation region. Binoculars reveal the four stars of the Trapezium cluster, which illuminate the nebula.

The Belt points up to the northwest towards **Taurus** (the Bull) and orange-tinted **Aldebaran** (α Tauri). Close to Aldebaran, there is a conspicuous 'V' of stars called the **Hyades** cluster. Aldebaran lies much closer to the Earth than the Hyades, it just so happens to share the same line of sight. Farther along, the same line from Orion passes below a bright cluster of stars, the **Pleiades**, or Seven Sisters. Even the smallest pair of binoculars reveals this as a beautiful group of bluish-white stars. The other conspicuous star in Taurus, **Elnath** (β Tauri), lies directly above Orion and forms part of a big 'V' with Aldebaran.

Running south from Orion is the long constellation of **Eridanus** (the River), which begins near Rigel in Orion and runs far south to end at **Achernar** (α Eridani). To the south of Orion is the constellation of **Canis Major** and several other constellations, including the oddly shaped **Carina**. The line of Orion's Belt also points southeast in the general direction of **Sirius** (α Canis Majoris), the brightest star. Almost due south of Sirius lies **Canopus** (α Carinae), the second brightest star in the sky.

Stellar birth: Bok globules

Stars are born in vast cool clouds (nebulae) of gas and dust in the spiral arms of the Milky Way. Our Sun formed 4.5 billion years ago, most likely with sibling stars. Over time, this stellar nursery dispersed, the stars evolved and migrated to other parts of the galaxy. Some stars end up in orbit around each other, becoming binary or multiple-star systems.

In the 1940s, Dutch astronomer Bart Bok noticed many small, very dark regions in a bright nebula. In January 1947, in collaboration with American astronomer Edith Reilly, Bok proposed these were dense, opaque, cool clouds of gas and dust, collapsing to form an embryonic star – a protostar. Light from these young stars is blocked by dust grains composed of silica and carbon compounds; the dust also acts as a shield protecting the stars from external radiation. They described these clouds, known as **Bok globules**, as an insect's cocoon, enshrouding the developing star. In 1990 (after Bok's death), infrared telescopes were able to image these dust clouds and hidden stars, thus confirming Bok's theory of star formation.

Bok globules are a type of dark molecular cloud, usually around 1 light-year across, and occur within huge, ionized hydrogen nebulae surrounding young supergiant stars in the spiral arms of our galaxy. The globules are between 2 and 50 times the mass of the Sun, and they are comprised mostly of molecular hydrogen, and around 1 per cent dust. Inside the globules, where the temperature is as low as –260°C, gravity takes over and pulls matter together into a dense core. Cloud collapse occurs when a threshold is reached called the Jeans mass; this depends on the temperature and density of the cloud, but is typically in the range of thousands of solar masses. As the gas collapses, gravitational potential energy converts into thermal energy and the core heats up. The hot fragments of the cloud become rotating spheres of gas, forming stellar embryos. A protostar is formed when the radiation pressure can counteract the gravitational collapse and the star enters a state of hydrostatic equilibrium. Some globules contain **Herbig-Haro objects**, which are bright areas produced when jets of gas

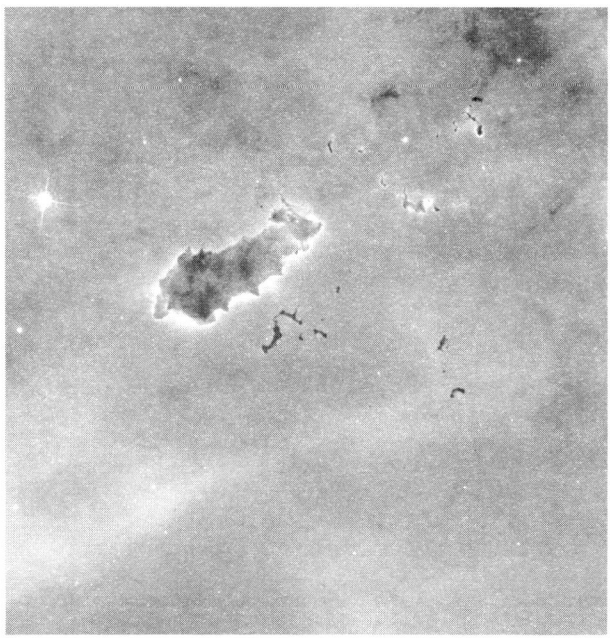

The Caterpillar, a Bok globule in the Carina Nebula, imaged by the HST in 2005.

from young stars collide with gas and dust at several hundred kilometres per second.

There are an estimated 6,000 molecular clouds in the Milky Way, each with more than 100,000 solar masses. The nearest site of star formation is the **Orion Nebula**, 1,300 light-years away and visible to the naked eye. Other well-known stellar nurseries are the **Carina Nebula**, 7,500 light-years from Earth, and the **Tarantula Nebula** in the **Large Magellanic Cloud**.

February

February – Introduction

Meteors

The **Centaurid** shower (which actually consists of two separate streams: the **α-** and **β-Centaurids**), whose radiants lie near α and β Centauri (Rigil Kentaurus and Hadar), respectively, continues in February, visible to southern observers. The shower reaches a low maximum of around 6 meteors per hour on 8 February, when the Moon is a bright waning gibbous rising after midnight. Another weak shower, the **γ-Normids**, begins to be active on 25 February, but the meteors are difficult to differentiate from sporadics. It peaks on 14 March, with a similar hourly rate to the Centaurids.

The Planets

Mercury is in **Capricornus** just after sunset; it moves into **Aquarius** and then **Pisces** in the final week of the month. It dims from mag. −1.2 to 1.5. On 18 February Mercury (mag. −0.6) sits only 0.1°N of the thin waxing crescent Moon, the day after it reaches eastern elongation. **Venus** is also present after sunset, at mag. −3.9 all month. It passes 4.7°S of Mercury on 26 February, a challenging observation due to the glow of the sunset. **Mars** starts in Capricornus with Mercury and moves slowly into Aquarius at the end of the month, at mag 1.1. **Jupiter** stays in **Gemini** and can be seen as soon as skies are dark (mag. −2.6 to −2.4). On 27 February it lies 4°S of the waxing gibbous Moon. **Saturn** is in Pisces and observable for a few hours after sunset. On 16 February it appears 1°S of **Neptune**, at mag 1.0 with Neptune at mag. 7.8. **Uranus** sits in **Taurus** from sunset until midnight (mag. 5.7) and ends its retrograde motion on 4 February. Neptune lies in Pisces with Saturn (mag. 7.8).

Oleg Kononenko, a Russian cosmonaut, has spent a total of 1,111 days in space. He broke the world record for the most time spent in space on 4 February 2024. He has conducted seven EVAs (extravehicular activity) or spacewalks – a total of 44.5 hours outside the **International Space Station (ISS)**.

Oleg Kononenko on EVA to examine the external hull of Soyuz MS-09, *standing on a Strela crane, on 11 December 2018.*

A team in Paris discovered a young ocean deep beneath the icy surface of Saturn's moon, **Mimas**. The ocean of water formed between 5–15 million years ago, and its presence was deduced from analysis of the moon's irregular orbit around Saturn. The moon is one of at least 274 confirmed moons, the largest being Titan with a diameter of 5,000 km. Mimas is only 400 km wide and its surface is heavily cratered. Results published in *Nature* in 2024 reveal the ocean to be present 20–30 km beneath the surface.

**Nicolaus Copernicus: The heliocentric model
of the Solar System**

In the fifteenth century it was widely believed the Earth was
at the centre of the Universe, a model first proposed by the
Greek philosopher Aristotle, and later by the Greek-Egyptian
mathematician and astronomer Ptolemy. To explain observed
motions of the planets visible to the naked eye – Mercury,
Venus, Mars, Jupiter and Saturn – Ptolemy suggested planets
moved in small circles called **epicycles** as they made their way
around the Earth in a sphere called a **deferent.** To complicate
matters further, the deferent of each planet rotated around
a point outside of the Earth, this point also rotated around
another point called an **equant**, and there was a different equant
for each planet.

The first recorded proposal of the heliocentric model of
the Solar System, in which the Earth and other planets orbit the
Sun, was made by Aristarchus of Samos in 250 BCE. Supporters
of the geocentric model found reasons to discredit this new
model; Earth at the centre of the cosmos remained the model
up until the mid-1500s, fully supported by the Christian
church. In 1545, Nicolaus Copernicus, a Polish mathematician,
astronomer and Catholic canon born on 19 February 1473,
published a book called *De
revolutionibus orbium coelestium*
(*On the revolutions of the Celestial
Spheres*). He proposed that the
Earth spins on its axis once
every 24 hours, not the stars in
the night sky. He explained that
the Moon does in fact orbit the
Earth; however, he argued that
the Sun must be at the centre
of the Solar System. This model
explained years of observations
of the planets without the need
for Ptolemy's complicated
epicycles and shifting orbital
spheres. Copernicus added

Nicolaus Copernicus, 1580

the theory that the stars were much farther away from the Solar System than originally believed.

It took a century or so for his theories to be widely accepted among the astronomical community. Copernicus' model was adjusted by Johannes Kepler; the circular orbits of the planets were found to be in fact elliptical (slightly squashed circles). The Catholic Church did not initially object to his theories; however, the book was condemned in 1616 – a response triggered by a dispute between the Church and the Italian

F

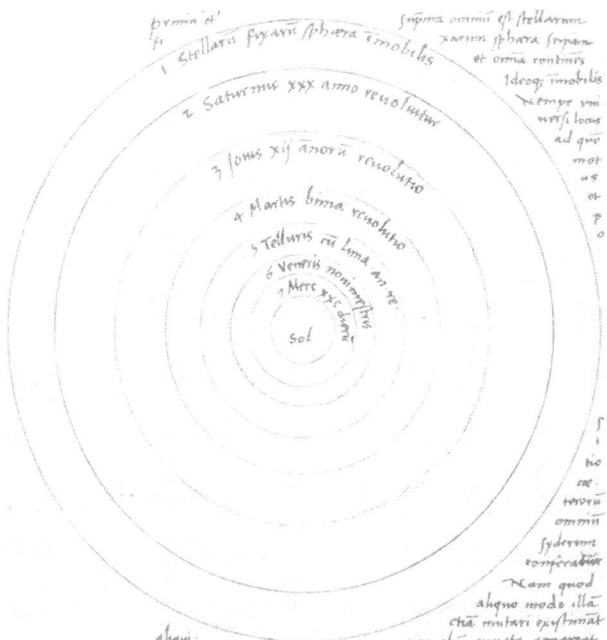

Copernicus' diagram of the heliocentric model for the Solar System, published in his book De revolutionibus orbium coelestium *(1473).*

astronomer and passionate heliocentric model supporter, Galileo Galilei. Galileo's telescopic observations of Jupiter's four largest moons and the phases of Venus provided evidence for the heliocentric model, and he published his drawings and results in March 1610. After further investigation of Copernicus' work, the Church banned his book and put Galileo under house arrest.

Copernicus' theory kickstarted the Scientific Revolution that swept Europe between the sixteenth and eighteenth centuries, transforming human knowledge and society.

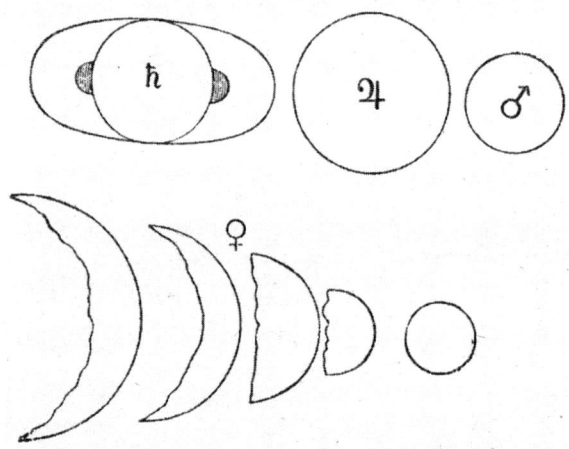

Galileo's drawings of the phases of Venus, evidence for the heliocentric theory.

Sunrise and sunset

City	Date	Sunrise	Sunset
Buenos Aires, Argentina			
	Feb. 01	09:14	23:00
	Feb. 28	09:40	22:32
Cape Town, South Africa			
	Feb. 01	04:08	17:52
	Feb. 28	04:33	17:24
London, UK			
	Feb. 01	07:39	16:50
	Feb. 28	06:48	17:39
Los Angeles, USA			
	Feb. 01	14:50	01:24
	Feb. 28	14:23	01:49
Nairobi, Kenya			
	Feb. 01	03:41	15:51
	Feb. 28	03:41	15:49
Sydney, Australia			
	Feb. 01	19:16	09:01
	Feb. 28	19:42	08:33
Tokyo, Japan			
	Feb. 01	21:41	08:08
	Feb. 28	21:13	08:35
Washington, DC, USA			
	Feb. 01	12:14	22:30
	Feb. 28	11:42	23:00
Wellington, New Zealand			
	Feb. 01	17:26	07:42
	Feb. 28	18:00	07:06

F

NB: the times given are in Universal Time (UT)

The Moon's phases and ages

Northern Hemisphere

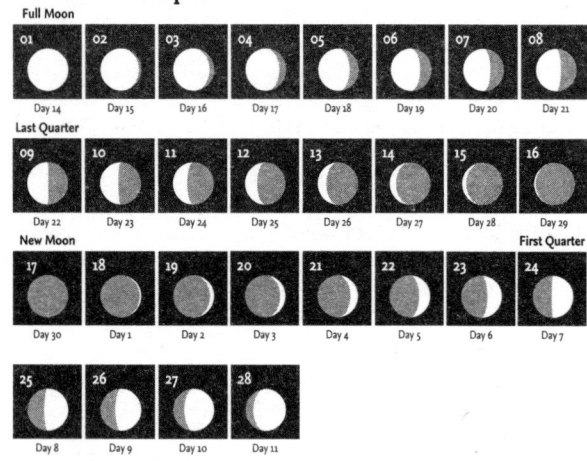

Southern Hemisphere

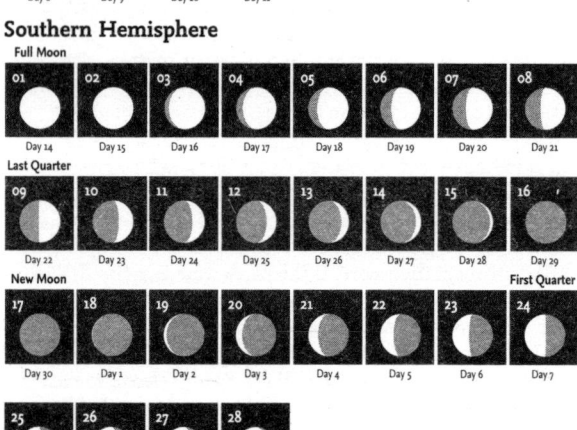

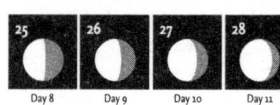

The Moon

The Moon in February

The month begins with a Full Moon, two days later **Regulus** lies 0.4°S of the waning gibbous Moon. On 7 February **Spica** lies 1.8°N of the Moon; the Moon reaches Last Quarter on 9 February. The red supergiant **Antares** can be seen 0.7°N of the waning crescent Moon on 11 February. A New Moon on 17 February is followed by a conjunction with **Mercury** (mag. −0.6) the following day close to the sunset, with the planet only 0.1°N of the thin sliver of waxing crescent Moon. **Saturn** (mag. 1.0) lies 4.6°S of the Moon on 19 February. On 24 February the **Pleiades** lie 1.2°S of the First Quarter Moon. On 27 February **Jupiter** (mag −2.5) will be placed 4.0°S of the waxing gibbous Moon along with **Pollux** 3.0°N of the Moon.

Black holes seed galaxy growth

Black holes are the end points of massive stars. They are described as black due to the fact that once light passes a boundary called the **event horizon** it is pulled in by gravity and cannot escape. Black holes come in a range of masses from a few times the mass of the Sun to billions of solar masses. These monster black holes are called **supermassive black holes (SMBH)** and are found in the hearts of galaxies, including our Milky Way. The SMBH in our galaxy, called Sagittarius A*, lies 26,000 light-years away and is 4.3 million solar masses. There may be up to 100 million stellar-mass black holes in the Milky Way alone, lying among 400 billion stars. There are an estimated trillion galaxies in the Universe, most of which have a SMBH sitting at their core. If the black hole is active, it can influence the rest of the galaxy through powerful, large-scale bipolar outflows, shaped by strong magnetic fields and swirling discs of matter around the black hole.

Observations from the **James Webb Space Telescope** reveal galaxies in the early Universe are brighter than expected. Analysis suggests that black holes and galaxies coexisted and influenced each other's evolution during the first 100 million years after the Big Bang, when the Universe was around 1 per cent of its current volume. The outflows from the SMBHs in these galaxies compressed huge gas clouds, igniting star formation – black holes kickstarted a starburst effect and were the seeds for early galaxy growth.

Carl Seyfert: Active galaxies

Between 1940 and 1942, American astronomer Carl Seyfert (1911–60) studied a number of spiral galaxies that had highly luminous compact cores. These galactic nuclei are bright across most of the electromagnetic spectrum, and are often most intense in the infrared region. Their visible and infrared spectra show very bright emission lines of hydrogen, helium, nitrogen and oxygen. The lines are broadened due to these elements moving at speeds of 500–4,000 kilometres per second. Since 1943, these galaxies have been categorized as Seyfert galaxies.

Seyfert galaxies are a type of active galaxy with a nucleus powered by a supermassive black hole, referred to as active galactic nuclei or AGN. They are classified according to their features: it is widely accepted that they all have a central black hole with a fast-moving accretion disc swirling around it. Beyond the disc is a dusty torus, or ring of gas and dust, which is opaque to visible light. The accretion disc and magnetic fields around the black hole funnel material along the rotational axis into two bipolar jets, which can extend thousands to millions of light-years, out into the rest of the galaxy and beyond. The orientation of these jets, accretion disc and torus relative to our line of sight defines the type of AGN observed. Quasars are another type of AGN, ubiquitous in the early Universe.

Seyfert galaxies constitute around 10 per cent of all galaxies in the Universe. The luminosity of their cores is comparable to the luminosity of a galaxy the size of the Milky Way. In contrast, the core of a quasar is brighter than the stars in the rest of the galaxy by a factor of 100 or more.

F

The Seyfert's Sextet, a group of galaxies 190 million light-years away in Serpens.

Messier 77, also called the Squid Galaxy, 47 million light-years away in Cetus, one of the first Seyfert galaxies classified.

Calendar for February

01	22:09	Full Moon
03	02:48	Regulus 0.4°S of the Moon
07	08:26	Spica 1.8°N of the Moon
08		α-Centaurids meteor shower maximum
09	12:43	Last Quarter Moon
10	16:52	Moon at apogee = 404,577 km
11	03:19	Antares 0.7°N of the Moon
17	12:01	New Moon
17	12:12	Annular solar eclipse (Antarctica), partial eclipse (south Argentina & Chile, south Africa)
18	23:03	Mercury (mag. −0.6) 0.1°N of the Moon. An occultation will be visible from Mexico, Australia and New Zealand.
19	17:39	Mercury at eastern elongation (18.1°E, mag. −0.6)
19	23:54	Saturn (mag. 1.0) 4.6°S of the Moon
24	02:43	Pleiades 1.2°S of the Moon
24	12:28	First Quarter Moon
24	23:18	Moon at perigee = 370,132 km
27	06:26	Jupiter (mag. −2.5) 4.0°S of the Moon
27	21:34	Pollux 3.0°N of the Moon

F

1–3 February: The Full Moon on 1 February lies in Cancer and passes through Leo as it wanes. It lies next to Regulus on 3 February (as seen from Sydney).

19 February: Saturn sits next to the waxing crescent Moon and Mercury (at greatest elongation) at dusk, low in the west (as seen from central USA).

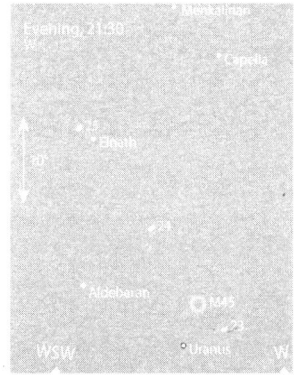

24 February: The First Quarter Moon and the Pleiades (M45). Uranus and Aldebaran are nearby (as seen from London).

27 February: The waxing gibbous Moon is next to Pollux (in Gemini) and Jupiter. Procyon is also visible (as seen from Sydney).

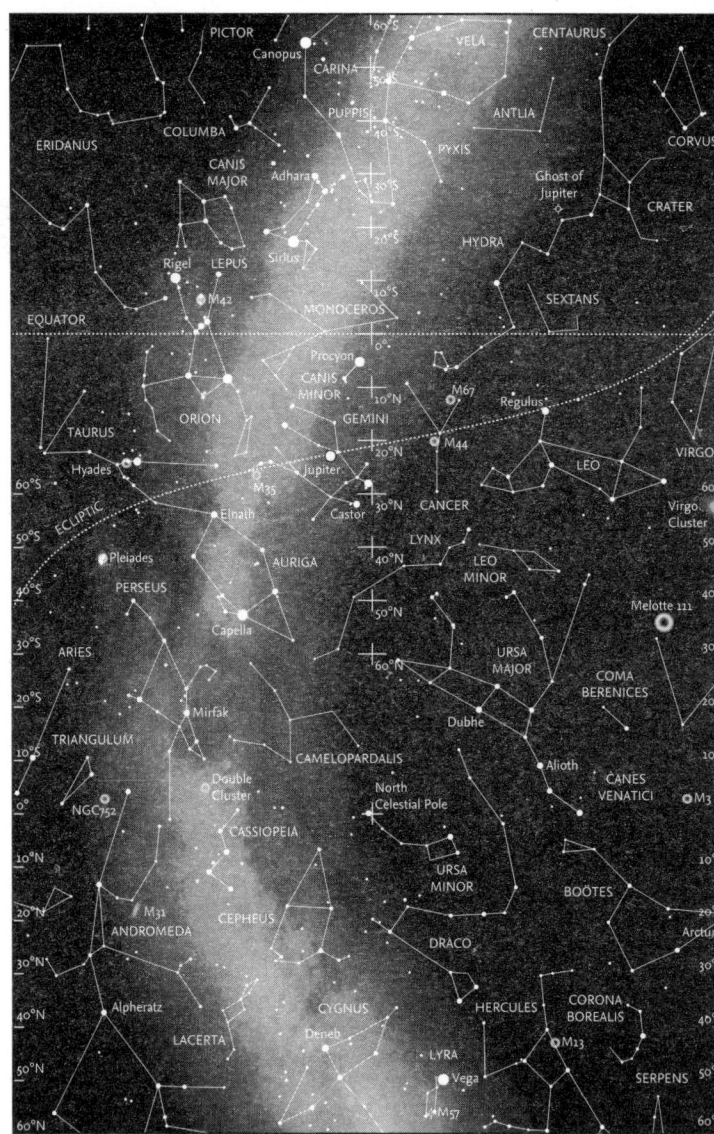

February – Looking North

The denser band of stars of the Milky Way, first observed through a telescope by Galileo Galilei, run through the northern and western sky. The band stretches from **Cygnus**, low on the northern horizon, through **Cassiopeia**, **Perseus** and **Auriga** and reaches the southern sky via **Gemini** and **Monoceros**. Although not as readily visible as the denser star clouds of the summer Milky Way, on a clear night so many stars may be seen that even a distinctive constellation such as Cassiopeia is not immediately obvious.

The 'base' of the constellation of **Cepheus** lies on the edge of the stars of the Milky Way, but the red supergiant star **Mu (μ) Cephei**, called the Garnet Star by William Herschel, remains readily visible with its striking red colour. The groups of stars known as the **Double Cluster** in Perseus (NGC 869 & NGC 884, often known as η and χ Persei), lying between Perseus and Cassiopeia, are well-placed for observation.

The head of **Draco** is now higher in the sky and easier to recognize. The blue supergiant **Deneb** (α Cygni), the brightest star in **Cygnus**, may just be visible almost due north at midnight and **Lyra**, with **Vega** (α Lyrae), may be seen at times. Most of the constellation of **Hercules** is visible, together with the distinctive circlet of **Corona Borealis** to its east. Later in the night the constellation of **Boötes** the herdsman, resembling a kite or diamond, clears the horizon. The orange-tinted red giant **Arcturus** (α Boötis) marks an apex. At magnitude −0.05, it is the brightest star in the northern hemisphere of the sky. **Coma Berenices** (Berenice's Hair) is now well above the horizon in the east. Within the constellation is a faint open star cluster of around 40 stars called Melotte 111, or the Coma Star Cluster.

On the other side of the sky, in the northwest, most of the constellation of **Andromeda** can still be seen, although **Alpheratz** (α Andromedae), the star that forms the northeastern corner of the Great Square of Pegasus – even though it is actually part of Andromeda – is approaching the horizon and becoming more difficult to detect.

An exoplanet over twice the diameter of the Earth, most likely made of rock, was discovered to orbit its host star once every 18.8 Earth-days. The planet, called **TOI-1437 b**, is thought to have a temperature of 1,200°C.

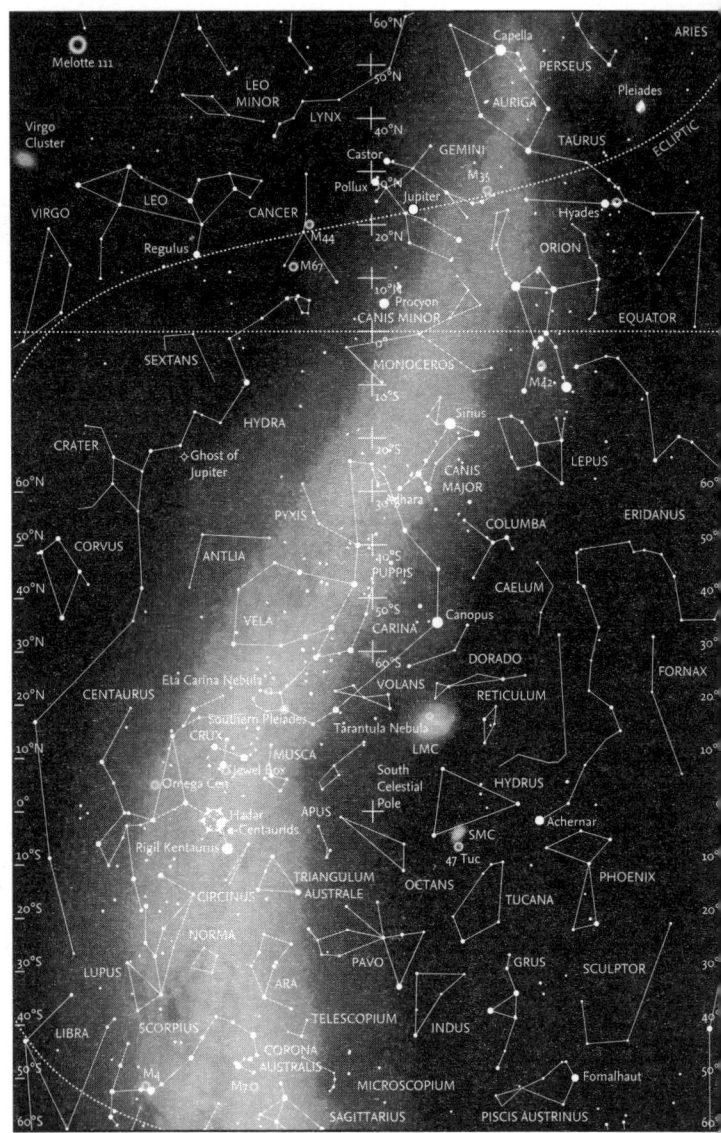

February – Looking South

Apart from **Orion**, the most prominent constellation visible this month is **Gemini**; the heads of the twins are marked by the bright stars **Castor** and **Pollux**, and the lines marking the bodies run southwest towards Orion. Castor (α Geminorum), the fainter star (mag. 1.9), is closer to the North Celestial Pole. Pollux (β Geminorum) is the brighter of the two (mag. 1.2) and it is occasionally occulted by the Moon. Castor is remarkable because it is actually a multiple system, consisting of no fewer than six individual stars arranged in three binary (paired) systems.

Orion's Belt points down to the southeast towards **Sirius**, the brightest star in the sky (at mag. –1.4) in the constellation of **Canis Major**. Forming an equilateral triangle with **Betelgeuse** in Orion and Sirius in Canis Major is **Procyon**, the brightest star in the small constellation of **Canis Minor**. Between Canis Major and Canis Minor is the faint constellation of **Monoceros**, which straddles the Milky Way. Directly east of Procyon is the highly distinctive asterism of six stars that form the 'head' of **Hydra**, the largest of all 88 constellations.

Canopus (α Carinae) is readily visible to observers at low northern latitudes and is close to the zenith for those in the far south. South of **Carina** and the neighbouring constellation of **Vela** lies the sprawling constellation of **Centaurus**, surrounding the distinctive constellation of **Crux**, the Southern Cross, which is on the horizon at 10°N. North of this, the False Cross, sometimes mistaken for Crux, consists of two stars from each of Vela (δ and κ Velorum) and Carina (ε and ι Carinae). The two brightest stars of Centaurus, **Rigil Kentaurus** (α Centauri) and **Hadar** (β Centauri), are slightly farther south. Northeast of Crux is the finest and brightest globular cluster in the sky, **Omega (ω) Centauri**, also known as NGC 5139. Discovered by Edmond Halley in 1677, it is estimated to contain about 10 million stars.

The **Large Magellanic Cloud** (LMC) is almost on the meridian, early in the night, to the west of Carina and the small constellation of **Volans**. Any optical aid (such as binoculars) will begin to show some of the remarkable structures within the LMC, including the great **Tarantula Nebula**, or **30 Doradus**, a site of active star formation.

Farther south and west lies ***Achernar*** (α Eridani), the bright star at the end of ***Eridanus***, the long straggling constellation that represents a river and that may now be traced all the way from where it begins near ***Rigel*** (β Orionis) in Orion.

Sirius, α Canis Majoris (α CMa), in the southern celestial hemisphere, is the brightest star in the sky at mag. −1.44.

March

March – Introduction

The Sun crosses the celestial equator on 20 March, at the vernal equinox, when day and night are of almost equal length, and the northern season of spring is considered to begin. The hours of daylight change most rapidly around the equinoxes in March and September. It is also in March that British Summer Time (BST) begins (on Sunday 29 March). In the rest of Europe, Daylight Saving Time is introduced on the same date.

The point at which the Sun crosses the celestial equator is known as the *First Point of Aries*, and originally lay in that constellation. Because of the phenomenon of precession, which changes the orientation of the Earth's axis in space, this point now actually lies in the constellation of *Pisces*, close to the border with *Aquarius*, lying near the star λ *Piscium*. It is slowly moving towards Aquarius at a rate of about one degree every 70 years.

The autumnal equinox (or spring equinox for those in the southern hemisphere), when the Sun crosses the celestial equator in the opposite direction (north to south), occurs in September (23 September in 2026). This intersection point lies in the large constellation of *Virgo* (see page 176). The First Point of Aries is sometimes known as the Cusp of Aries, and the equinox in September as the Cusp of Libra.

Meteors

The meteor shower in March for southern observers is the *γ-Normids*, which have a low rate, and are thus difficult to differentiate from sporadics. However, they exhibit a very sharp peak a day or so on either side of maximum on 14 March, when the Moon is waning crescent. The faint constellation of *Norma* rises early in the night, but most meteors are likely to be seen (away from the radiant) in the hours after midnight, before the Moon rises.

The planets

Mercury lies in *Pisces* trailing the Sun at the beginning of the month. After the first week it leads the Sun, eventually appearing early in the dawn sky in *Aquarius*, dimming from mag. 2.0 then brightening to 0.4. On 15 March it joins *Mars* (mag. 1.2) lying 3.4°N. *Venus* enters Pisces, dipping into *Cetus*

and back as the month progresses and then reaching **Aries** at
mag. −3.9. It passes 1°N of **Saturn** (mag. 0.9) on 8 March, visible
just after sunset. **Mars** lies in Aquarius for the duration of
the month (mag. 1.1 to 1.2). **Jupiter** continues to move slowly
through **Gemini** (mag. −2.4 to −2.2). On 11 March it ends
retrograde motion and moves eastward. Saturn is in Pisces,
setting shortly after sunset and swinging into the dawn sky after
25 March at mag. 0.9. **Uranus** continues in **Taurus**, dimming
from mag. 5.7 to 5.8 and **Neptune** in Pisces approaches the Sun
and from 22 March moves into the dawn sky (mag. 7.8 to 7.9).

M

Sunrise and sunset

City	Date	Sunrise	Sunset
Buenos Aires, Argentina			
	Mar. 01	09:41	22:30
	Mar. 31	10:06	21:50
Cape Town, South Africa			
	Mar. 01	04:34	17:23
	Mar. 31	04:58	16:43
London, UK			
	Mar. 01	06:46	17:40
	Mar. 31	05:38	18:32
Los Angeles, USA			
	Mar. 01	14:21	01:49
	Mar. 31	13:41	02:13
Nairobi, Kenya			
	Mar. 01	03:41	15:49
	Mar. 31	03:34	15:40
Sydney, Australia			
	Mar. 01	19:43	08:32
	Mar. 31	20:06	07:52
Tokyo, Japan			
	Mar. 01	21:12	08:36
	Mar. 31	20:30	09:01
Washington, DC, USA			
	Mar. 01	11:40	23:01
	Mar. 31	10:54	23:31
Wellington, New Zealand			
	Mar. 01	18:01	07:05
	Mar. 31	18:35	06:15

NB: the times given are in Universal Time (UT)

The Moon's phases and ages

Northern Hemisphere

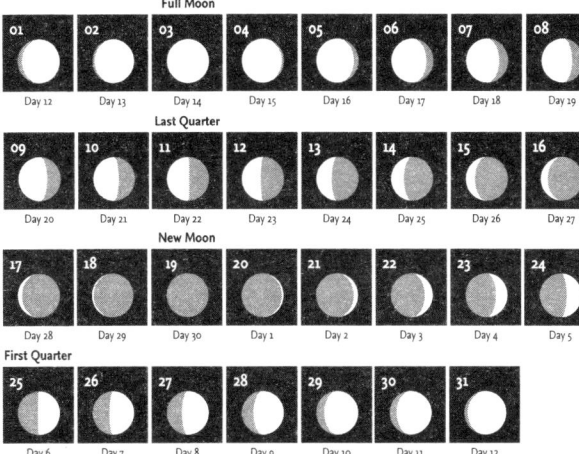

Southern Hemisphere

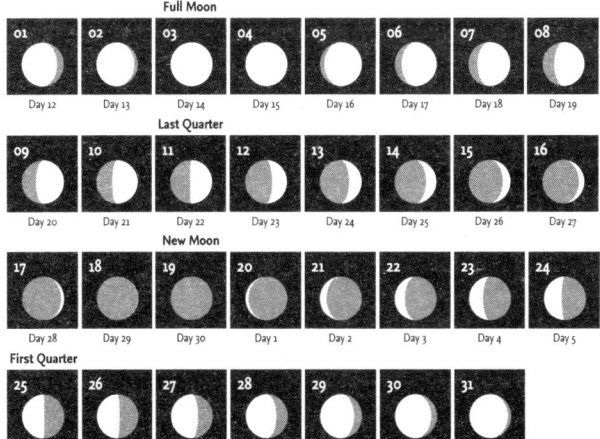

The Moon

The Moon in March

On 2 March the waxing gibbous Moon is in *Leo*, 0.4°N of *Regulus*. A Full Moon occurs on 3 March. *Spica* lies 1.8°N of the waning gibbous Moon three days later. On 10 March *Antares* can be seen 0.7°N of the Moon. A day later the Moon is Last Quarter. On 17 March *Mercury* (mag. 1.7) is 2.0°N of the waning crescent Moon and *Mars* (mag. 1.2) is 1.5°S of the Moon. The New Moon appears on 19 March. A day later *Venus* (mag. −3.9) lies 4.6°S of the very thin crescent Moon. On 23 March, two days before it reaches First Quarter, the Moon sits 1.1°N of the *Pleiades*. On 26 March *Jupiter* (mag −2.2) passes 3.9°S of the waxing gibbous Moon. The following day *Pollux* appears 3.0°N of the Moon. The bright Moon returns to *Regulus* on 29 March, lying 0.4°N of the brightest star in *Leo*.

Ancient stars

A Population II star that was born in another galaxy has been found in our galaxy. Research at the University of Chicago published in March 2024 revealed the existence of this ancient star from the previous generation to our Sun. The star will provide much-needed clues as to what Population III stars may have been like as their material was used to form Population II stars. It is thought there are only 100,000 Population II stars in our galaxy, less than a millionth of the total number of stars present. Analysis of the chemical composition of the outer layers of these stars uncovers the chemistry of their birthplace billions of years ago. Older generations of stars have fewer heavier elements; the star analysed by the team was found to have significantly less carbon than iron, in contrast to younger Population I stars like the Sun. The ancient star lies in the Large Magellanic Cloud, a satellite dwarf galaxy of the Milky Way, once thought to be a separate galaxy captured by the gravity of our galaxy.

Astronomers have yet to observe the oldest generation of stars. This stellar archaeology fills in gaps in our theories of how stars shaped the cosmos over time, eventually forming planetary systems – and in our Solar System, life.

Lunar eclipse of 3 March
There are two total lunar eclipses in 2026. The first takes place on 3 March and it will be visible from eastern Asia and Australia, as well as parts of North and South America. During a total lunar eclipse the Moon moves eastwards into the Earth's penumbra, then moves behind the Earth into the darkest region of the Earth's shadow – the umbra – eventually creeping back out into the penumbra and into the sunlight. On 3 March maximum eclipse occurs at 11:34 UT. The Moon will spend a total of 58 minutes immersed in the darkest shadow of the Earth.

During this time the Moon is still visible but it will appear a reddish-brown colour, as it is only illuminated by refracted sunlight which passes through the Earth's atmosphere and bends towards the lunar surface. Bluer hues are scattered outwards by the Earth's atmosphere, leaving behind redder light that eventually reaches the Moon.

M

The Universe began 13.8 billion years ago and it is thought that the first stars came into existence 100 to 200 million years after the Big Bang. These are called Population III stars. Our Sun is a Population I star, a grandchild of the first cosmic stellar generation. In May 2024, astronomers at the University of Arizona using NASA's **James Webb Space Telescope** discovered a large and extraordinarily bright galaxy present when the Universe was only 290 million years old, during its cosmic dawn.

Calendar for March

02	12:00	Regulus 0.4°S of the Moon
03	11:34	Total lunar eclipse, visible from eastern Asia, Australia, parts of North and South America
03	11:38	Full Moon
06	17:24	Spica 1.8°N of the Moon
10	11:32	Antares 0.7°N of the Moon. An occultation will be visible from Antarctica and New Zealand.
10	13:43	Moon at apogee = 404,385 km
11	09:39	Last Quarter Moon
14		γ-Normid meteor shower maximum
15	19:34	Mercury (mag. 2.7) 3.4°N of Mars (mag. 1.2)
17	14:07	Mercury (mag. 1.7) 2.0°N of the Moon
17	21:51	Mars (mag. 1.2) 1.5°S of the Moon
19	01:23	New Moon
20	12:39	Venus (mag. –3.9) 4.6°S of the Moon
20	14:46	Vernal Equinox
22	11:40	Moon at perigee = 366,858 km
23	08:32	Pleiades 1.1°S of the Moon
25	19:18	First Quarter Moon
26	12:13	Jupiter (mag. –2.2) 3.9°S of the Moon
27	03:18	Pollux 3.0°N of the Moon
29	19:00	Regulus 0.4°S of the Moon. An occultation will be visible from Asia, Africa, Europe (including London) and western Russia.

M

2–3 March: The Moon passes close to Regulus on 2 March and becomes Full a day later (as seen from Sydney).

20 March: Venus lies south of the thin waxing crescent Moon in Pisces (as seen from central USA).

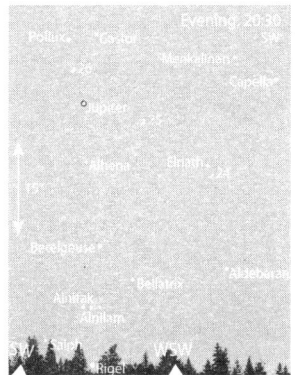

25 March: The First Quarter Moon is due southwest of Jupiter in Gemini. Pollux, Castor, Capella, Betelgeuse and Aldebaran are close by (as seen from London).

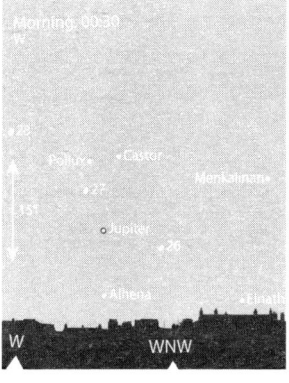

27 March: Pollux and the waxing gibbous Moon together with Jupiter in Gemini, low in the west (as seen from London).

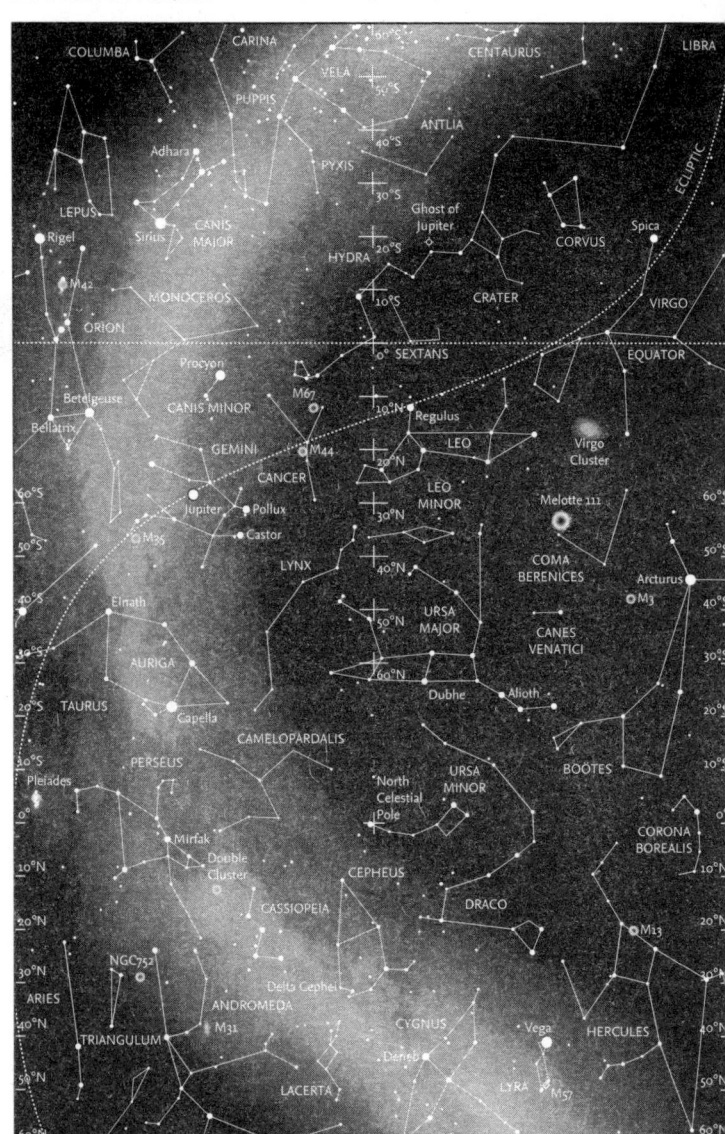

March – Looking North

At this time of year, **Ursa Major** with the distinctive asterism of the Plough (or Big Dipper) is now 'upside down' and high in the sky for anyone north of the equator. At 30°S, even the seven stars making up the main, easily recognized portion of the constellation are too low to be visible. **Auriga**, with brilliant **Capella** (α Aurigae), is also very high on the opposite side of the meridian. The constellation of **Perseus** lies between it and **Andromeda** on the western side of the sky.

Ursa Minor, also with seven main stars, one of which is **Polaris**, the Pole Star, and the long constellation of **Draco** that winds around the Pole, are readily visible for anyone in the northern hemisphere; although Polaris is right on the horizon for anyone at the equator, and thus always lost to sight. Early in the month, the constellation of **Cepheus** lies almost due north, with the distinctive 'W' of **Cassiopeia** to its west. Cepheus lies across the border of the Milky Way and one star stands out because of its deep red colour. This is **μ Cephei**, also known as the Garnet Star and Erakis. It is a truly gigantic star, with a diameter 1,400 times that of the Sun, and if placed in the Solar System the surface of the star would extend beyond the orbit of Jupiter. **Betelgeuse**, in **Orion**, is also a red supergiant; however, it is one third of the size of μ Cephei.

Another famous and very important star in Cepheus is **δ Cephei**, one of the class of variable stars known as Cepheids. These giant stars show a regular variation in their luminosity; the greater their intrinsic brightness, the longer the period of these changes. This can be used to determine their distance. Cepheid variables are the first major 'rung' in the cosmic distance ladder.

In the east lies the top of **Boötes**. Farther to the south, most of **Hercules** and the Keystone shape that forms the major portion of the body is visible.

Below Cepheus to the east it may be possible for observers at 50°N to catch a glimpse of **Deneb** (α Cygni), just above the horizon. Towards the northeast, **Vega** (α Lyrae) is marginally higher in the sky. Most of the time they will be lost in the extinction that occurs at such low altitudes.

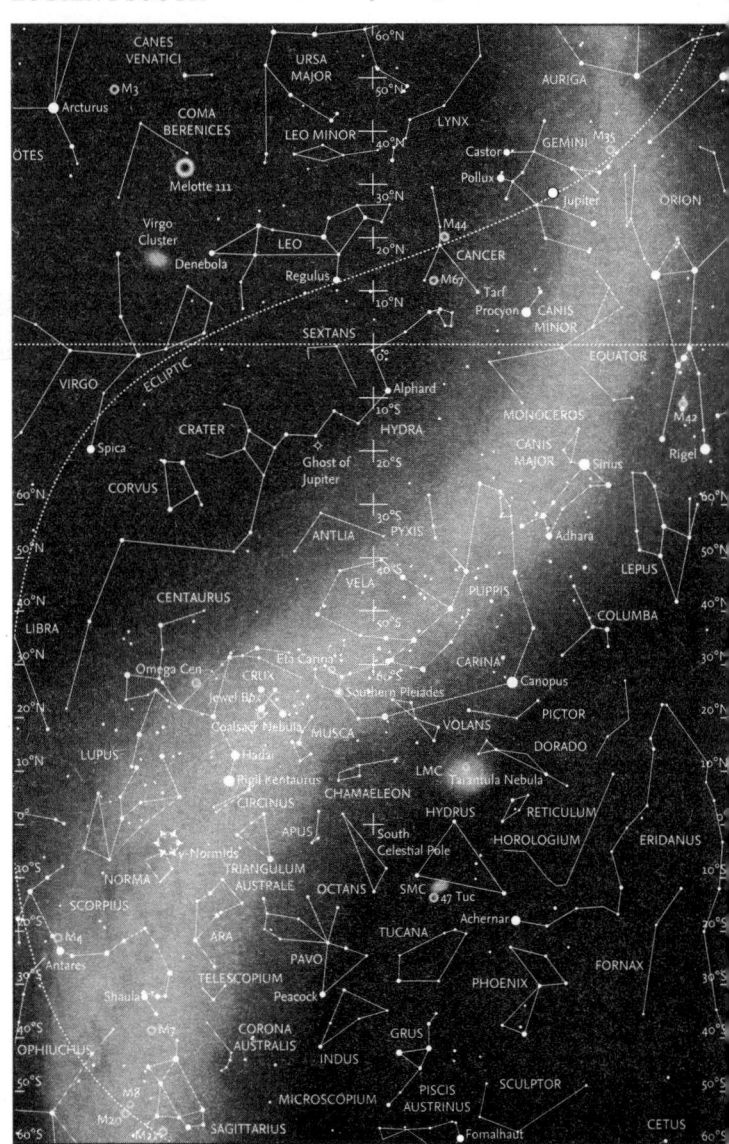

March – Looking South

Lying between the constellations of **Gemini** in the west and **Leo** in the east, and fairly high in the sky above the head of **Hydra**, is the zodiacal constellation of **Cancer** the crab. Its brightest star is β Cancri (also called **Tarf**), an orange giant around 50 times the size of the Sun. Near the centre of the constellation lies an open cluster, M44 or **Praesepe** (the Manger but also known as the Beehive). On a clear night this cluster of around 1,000 stars, known since antiquity, is just a hazy spot to the naked eye, but appears in binoculars as a group of dozens of individual stars.

Also prominent in March for observers in the northern hemisphere is the constellation of Leo, with the 'backward question mark' (or Sickle) of bright stars forming the head of the mythological lion. **Regulus** (α Leonis) – the 'dot' of the 'question mark' or the handle of the sickle and the brightest star in Leo – lies very close to the ecliptic and is one of the few first-magnitude stars that may be occulted by the Moon. **Denebola** (β Leonis), a variable star, marks the tail of the lion.

To the west of Leo is the constellation of Cancer, and still farther away from the meridian, the far more striking constellation of Gemini, with the bright stars **Castor** (α Geminorum) and **Pollux** (β Geminorum). The constellation straddles the ecliptic, and Pollux may sometimes be occulted by the Moon (as may Regulus). Castor is remarkable in that even a fairly small telescope will show it as consisting of three stars (two fairly bright, and one fainter). However, more detailed investigation reveals that each of those stars is actually a double, so the whole system consists of no fewer than six stars.

Below Cancer is the very distinctive asterism of the Head of Hydra, consisting of five (or six) stars, that is the western end of the long constellation of **Hydra**, the largest constellation in the sky, that runs far towards the east, roughly parallel to the ecliptic. **Alphard** (α Hydrae) is south, and slightly to the west of Regulus in Leo, and is relatively easy to recognize as it is the only fairly bright star in that region of the sky. North of Hydra and between it and the ecliptic and the constellation of **Virgo** are the two constellations of **Crater** and **Corvus**. Farther west, the small constellation of **Sextans** lies between Hydra and Leo.

Farther south, the Milky Way runs diagonally across the sky and the constellation of **Vela** straddles the meridian. Slightly farther south is the constellation of **Carina**, with, to the west, brilliant **Canopus** (α Carinae), which lies below the constellation of **Puppis**, which is itself between Vela and **Canis Major** in the west. **Crux** (the Southern Cross) is southeast of Carina and the two principal stars of **Centaurus**, **Rigil Kentaurus** (α Centauri) and **Hadar** (β Centauri). The **Large Magellanic Cloud** (LMC) lies west of these stars, on the other side of the meridian.

Scorpius is beginning to become visible in the eastern sky, including the Cat's Eyes, the pair of stars **Shaula** and **Lesath** (λ and υ Scorpii, respectively) at the end of the 'sting'. Above Scorpius, the whole of the constellation of **Lupus** is now easy to see. The magnificent globular cluster of **Omega Centauri** lies northeast of **Crux**. The **Coalsack Nebula** (a dark nebula of interstellar dust obscuring starlight) and the denser region of the Milky Way in **Carina**, together with the **Eta Carinae Nebula** (containing massive stars considerably hotter than the Sun) and the **Southern Pleiades**, are well placed for observation. The **False Cross** on the **Carina/Vela** border is high in the sky, between the South Celestial Pole and the zenith. **Achernar** (α Eridani), the **Small Magellanic Cloud** (SMC) and **47 Tucanae** are considerably lower in the sky. **Peacock** (α Pavonis), a binary star system in the constellation **Pavo**, skims the horizon, just east of south.

Maarten Schmidt: Quasars – super bright ancient galaxies
In 1963 Maarten Schmidt, a Dutch astronomer at the Palomar Observatory in California, analysed the light from an object 2.5 billion light-years away. He calculated it was 4 trillion times brighter than the Sun: brighter than the whole Milky Way, and the brightest object yet discovered. The emission lines matched those of a galaxy, and they indicated a recession velocity of around 160 million km per hour. He called it a quasi-stellar radio source, shortened to **quasar**. Quasars belong to the early Universe – we see them as they were billions of years into the past. The object he observed, **3C 273**, was first discovered in 1959.

Quasar 3C 273 imaged by the Hubble Space Telescope, 2013.

Since his discovery, hundreds of quasars have been found, some with luminosities 100 times that of the Milky Way. A supermassive black hole in the centre feeding off nearby matter pulls in an accretion disc of particles moving at high velocities, which increase closer to the black hole, approaching the speed of light. These in turn radiate electromagnetic energy across a range of wavelengths, contributing to the high brightnesses observed.

The rising and setting of the Moon

Our ancestors used the Moon to keep track of time over the course of a month. The first calendars were lunar calendars tracking the lunar phase cycle over a period of 29.5 days. In reality the Moon takes just over 27 days to orbit the Earth. The phase cycle is longer because it takes a few extra days for the Earth, Moon and Sun to have returned to their original positions for the start of the next phase cycle, at the New Moon.

The Moon rises and sets like the Sun and most of the stars because of the 24-hour spin of the Earth. As the Moon makes its way around the Earth, the time of moonrise becomes progressively later. When the Moon is New or a thin crescent it is predominantly present during daylight, and it is highest in the sky at midday; two weeks later when the Moon is Full it rises in the early evening, and reaches the meridian at midnight. As the Moon rises or sets it is redder in colour due to the same effect seen at sunrise and sunset. Earth's atmosphere scatters blue components of reflected sunlight from the Moon in all directions and reflected red light passes through unfiltered, reaching our eyes – we see a red Moon.

The Moon is tidally locked with the Earth; we only ever see one hemisphere of the Moon, the near side. The spin of the Earth is slowing down. It would take another 50 billion years for one hemisphere of the Earth to be permanently locked with the Moon, in which case the Moon would be visible from only one-half of the Earth. However, this stage may not ever be reached: in around 4 billion years' time the Sun will expand into a red giant and engulf Mercury and Venus, and its approaching surface will scorch the Earth and Moon.

Moonset over the Mojave Desert, California, USA, 2015.

April

April – Introduction

Daylight Saving Time ends where used in Australia and in New Zealand on Sunday 5 April.

Meteors

For northern hemisphere observers, the **Lyrids** (often known as the **April Lyrids**) begins on 16 April and peaks on 22 April, a few days before the First Quarter Moon. The best time to look for the meteors will be after the Moon has set from 02:00 onwards, at this time the Earth will have turned towards the cloud of debris and it will be easier to spot them in the sky. There is another stronger shower, the **η-Aquariids**, that begins on 19 April and peaks on 6 May.

For southern hemisphere observers the **π-Puppid** shower begins on 15 April, however, because there is a First Quarter Moon coinciding with the shower maximum on 24 April, moonlight will interfere with observations. The hourly rate is variable but the meteors tend to be faint. The parent body is the comet 26P/Grigg-Skjellerup.

The planets

Mercury moves from **Aquarius** to **Pisces**, brightening from mag. 0.4 to −0.7. On 3 April Mercury reaches western elongation. Mercury (mag. −0.2) lies 1.7°S of **Mars** (mag. 1.2) and 0.5°S of **Saturn** (mag. 0.9) on 20 April, low on the eastern horizon. **Venus** begins the month in **Aries** and moves into **Taurus** in the latter part of the month. On 8 April Venus (mag. −3.9) sits 4.5°N of **(1) Ceres** (mag. 9.0) at dusk and on 24 April it passes within 1°N of **Uranus** (mag. 5.8). Mars moves from Aquarius to Pisces and fluctuates between mag. 1.2 and 1.3. On 19 April it passes 1.2°N of Saturn (mag. 0.9). **Jupiter** stays in **Gemini**, dimming from mag. −2.2 to −2.0. Saturn remains in Pisces and leads the sunrise, pulling away from the Sun as the month progresses (mag. 0.9). Uranus lies in Taurus at mag. 5.8 all month. **Neptune** is in Pisces low in the east before sunrise (mag. 7.8).

The volcanic Jovian moon Io has a lava lake, as observed by NASA's **Juno** spacecraft in December 2023 and February 2024. *Juno* flew within 1,500 km of the surface and captured the cooling lava lake called Loki Patera, which is 200 km long. It was also revealed that parts of the moon's surface are as smooth as glass, similar to obsidian glass at volcanic sites on Earth.

A

Sunrise and sunset

City	Date	Sunrise	Sunset
Buenos Aires, Argentina			
	Apr. 01	10:06	21:48
	Apr. 30	10:29	21:13
Cape Town, South Africa			
	Apr. 01	04:59	16:42
	Apr. 30	05:20	16:07
London, UK			
	Apr. 01	05:36	18:34
	Apr. 30	04:34	19:22
Los Angeles, USA			
	Apr. 01	13:40	02:14
	Apr. 30	13:05	02:36
Nairobi, Kenya			
	Apr. 01	03:34	15:39
	Apr. 30	03:28	15:32
Sydney, Australia			
	Apr. 01	20:07	07:51
	Apr. 30	20:29	07:16
Tokyo, Japan			
	Apr. 01	20:28	09:02
	Apr. 30	19:51	09:26
Washington, DC, USA			
	Apr. 01	10:53	23:32
	Apr. 30	10:11	24:00
Wellington, New Zealand			
	Apr. 01	18:36	06:13
	Apr. 30	19:07	05:29

NB: the times given are in Universal Time (UT)

The Moon's phases and ages

Northern Hemisphere

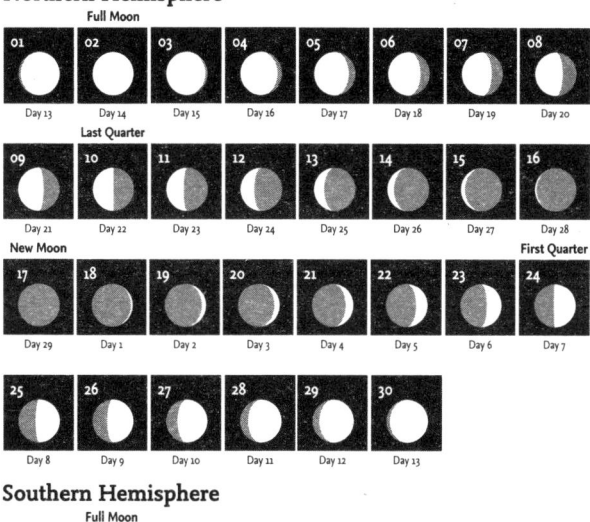

Southern Hemisphere

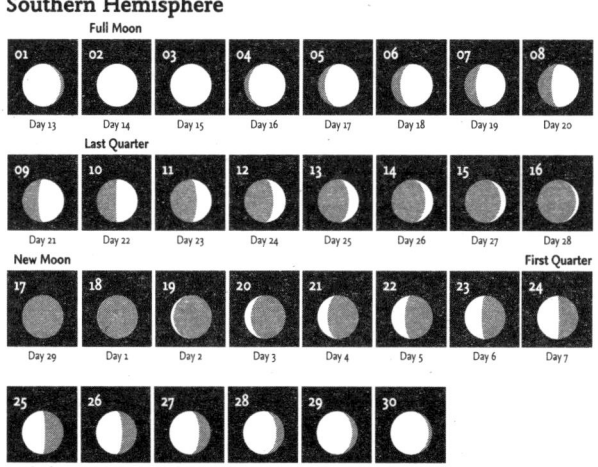

The Moon

The Moon in April

The Moon is Full on 2 April and as it begins to wane it will pass 1.8°S of **Spica**. On 6 April **Antares** and the waning gibbous Moon will be 0.6° apart, visible a few hours before sunrise. The Last Quarter Moon appears on 10 April. **Mars** (mag. 1.2) and the thin waning crescent Moon are 3.7° apart on 16 April. The following day brings a New Moon. The three-day-old waxing crescent Moon on 19 April lies 4.8°N of **Venus** (mag. –3.9) and the **Pleiades** cluster sits 1.0°S of the Moon. On 22 April **Jupiter** (mag. –2.1) can be seen 3.6°S of the Moon. The following day the Moon passes 3.2°S of **Pollux**. On 24 April the Moon is First Quarter. On 26 April the Moon approaches **Regulus**, sitting 0.2°N of the brightest star in **Leo**. The month ends with **Spica** and the waxing gibbous Moon sitting 1.8° apart.

Moon Standard Time

NASA is collaborating with other organizations to establish a Coordinated Lunar Time (LTC) by the end of 2026. In a process analogous to the way Coordinated Universal Time (UTC) is set, atomic clocks will be used. Due to the weaker gravity on the Moon, atomic clocks on the lunar surface will tick faster by a total of 56 microseconds (56 millionths of a second) per day, equivalent to 1 second every 46 Earth-years. In contrast, human reaction time is around a tenth of a second. This may seem like a negligible difference; however, if it isn't corrected for, it can cause problems when running machines on the Moon, such as spacecraft docking. If machines are running on different clocks, each will calculate a different approach distance, and this could result in a collision.

Lisa Kaltenegger: Astrobiologist

Austrian astrobiologist Lisa Kaltenegger hunts for extrasolar planets, or exoplanets for short – these are planets that orbit other stars. She is director of the Carl Sagan Institute at Cornell University, New York, named after the American planetary scientist who conducted research on the possibility of extraterrestrial life. Kaltenegger reported in *Nature* in 2021 that 1,715 stars within

325 light-years of Earth have had a direct line of sight to Earth transiting the Sun since the dawn of civilization (around 5,000 years ago). Radio waves leaking out into space over the past 100 years have touched over 75 of these stars (and potentially planetary systems). She published her book *Alien Earths* in April 2024, and has an asteroid named after her, called **7734 Kaltenegger**.

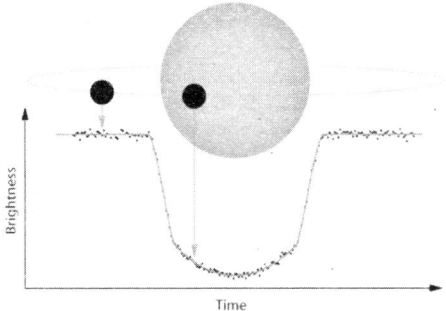

Periodic dips in the brightness of a star indicate the existence of transiting planets.

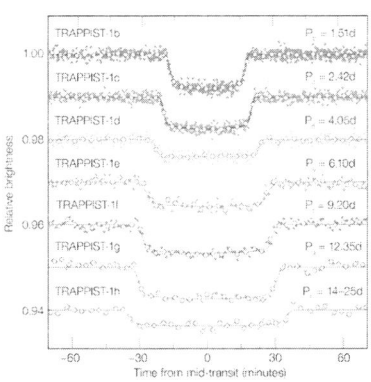

Light curves of the TRAPPIST-1 planetary system, a red dwarf star with at least seven exoplanets, lying 41 light-years away. Data taken by the NASA Spitzer Space Telescope, 2017.

Calendar for April

02	02:12	Full Moon
03	01:32	Spica 1.8°N of the Moon
03	22:33	Mercury at western elongation (27.8°W, mag. 0.4)
06	19:21	Antares 0.6°N of the Moon. An occultation will be visible from Antarctica, Madagascar, French Southern Territories and Mauritius.
07	08:32	Moon at apogee = 404,974 km
10	04:52	Last Quarter Moon
16	00:45	Mars (mag. 1.2) 3.7°S of the Moon
17	11:52	New Moon
19	06:57	Moon at perigee = 361,631 km
19	08:49	Venus (mag. −3.9) 4.8°S of the Moon
19	16:28	Pleiades 1.0°S of the Moon
19	17:36	Mars (mag. 1.2) 1.2°N of Saturn (mag. 0.8)
20	00:00	Mercury (mag. −0.2) 1.7°S of Mars (mag. 1.2)
20	08:03	Mercury (mag. −0.2) 0.5°S of Saturn (mag. 0.8)
22		Lyrid meteor shower maximum
22	22:06	Jupiter (mag. −2.1) 3.6°S of the Moon
23	08:59	Pollux 3.2°N of the Moon
24		π-Puppid meteor shower maximum
24	02:32	First Quarter Moon
24	04:17	Venus (mag. −3.9) 3.4°S of the Pleiades
26	00:37	Regulus 0.2°S of the Moon. An occultation will be visible from Brazil, the eastern contiguous United States, Colombia and Venezuela.
30	08:17	Spica 1.8°N of the Moon

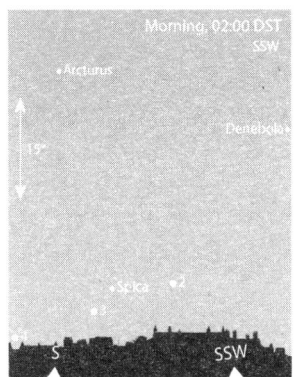

2–3 April: *The Full Moon moves through Virgo and passes close to Spica the day after. Arcturus and Denebola are in the same region of the sky (as seen from central USA).*

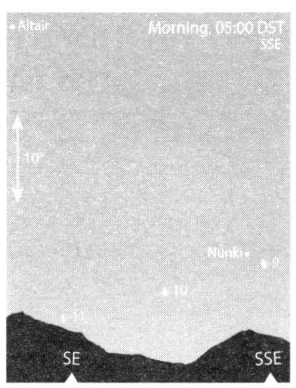

10 April: *Last Quarter Moon in Sagittarius (as seen from central USA).*

19 April: *Venus and the waxing crescent Moon in Taurus shortly after sunset (as seen from London).*

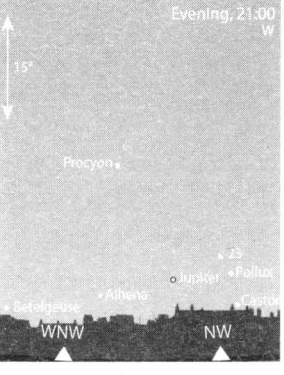

23 April: *The waxing crescent Moon moving into Gemini and approaching Jupiter, Pollux and Castor. Procyon is nearby (as seen from Sydney).*

A

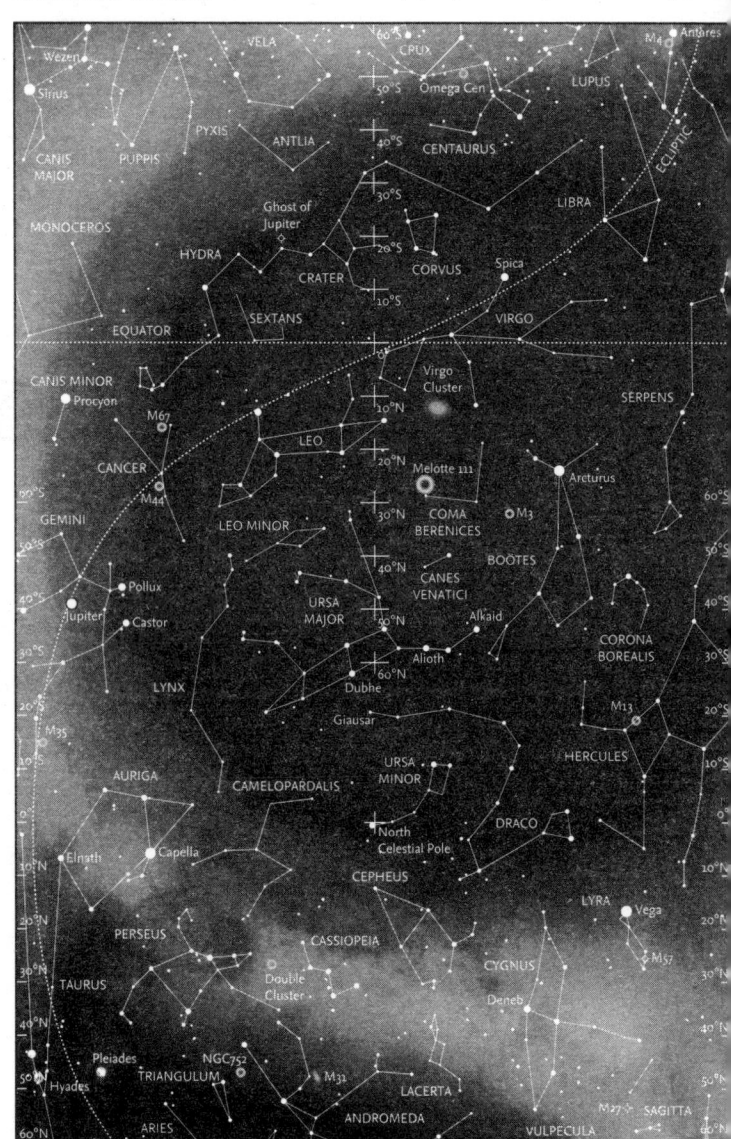

April – Looking North

Cygnus and the brighter regions of the Milky Way are now becoming visible, running more-or-less parallel with the horizon in the early part of the night. Rising in the northeast is the small constellation of *Lyra* and the distinctive Keystone of *Hercules* above it. This asterism is very useful for locating the globular cluster M13, which lies on one side of the quadrilateral. Some of the stars are 12 to 13 billion years old. *Cepheus* is beginning to climb higher and the winding constellation of *Draco* weaves its way from the four stars that mark its 'head', on the border with Hercules, to end at *Giausar* (λ Draconis) between *Polaris* (α Ursae Minoris) and the Pointers, *Dubhe* and *Merak* (α and β Ursae Majoris, respectively). *Ursa Major* is 'upside down' high overhead, near the zenith. *Cassiopeia* has now swung round, and is almost on the meridian to the north, below *Polaris* and northwest of *Ursa Minor*.

The constellation of *Gemini* stands almost vertically in the west. *Auriga* is still clearly seen in the northwest with its prominent star *Capella*, but, by the end of the month, the southern portion of *Perseus* is starting to dip below the northern horizon. The very faint constellation of *Camelopardalis* lies in the northwest between Polaris and the constellations of Auriga and Perseus.

A

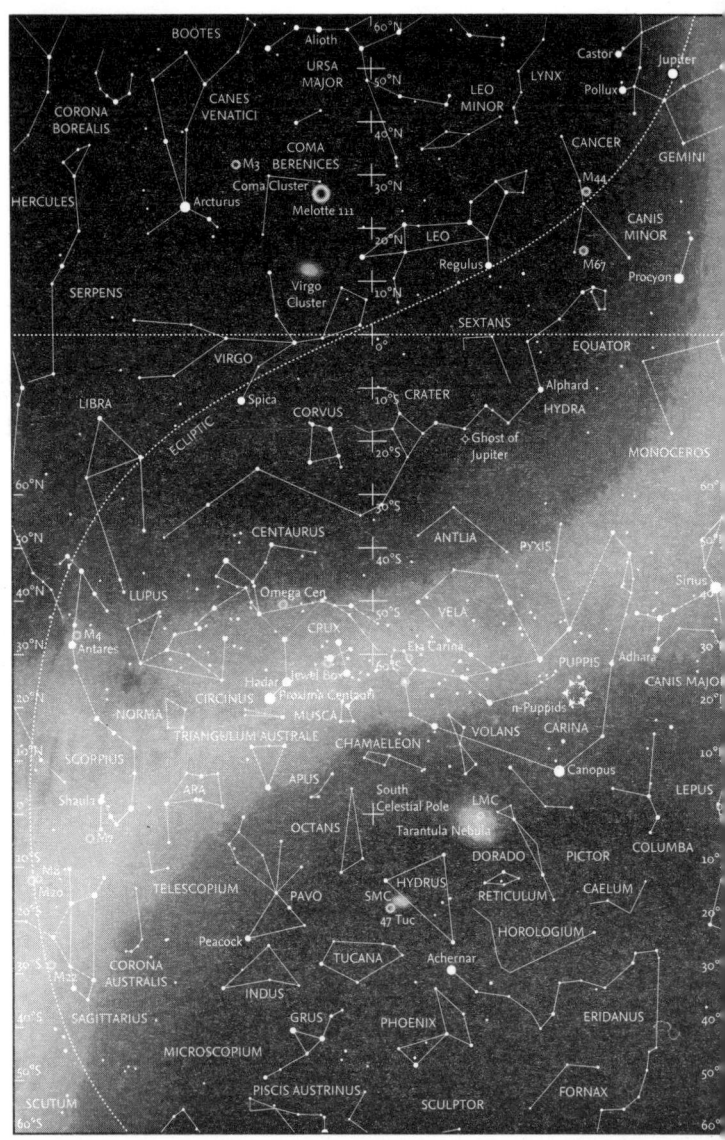

April – Looking South

Leo is the most prominent constellation in the southern sky in April and vaguely looks like the creature after which it is named. **Gemini**, with **Castor** and **Pollux**, remains clearly visible in the west, and **Cancer** lies between the two constellations. To the east of Leo, the whole of **Virgo**, with **Spica** (α Virginis) its brightest star, is well clear of the horizon. **Hydra** runs beneath both constellations, with **Alphard** (α Hydrae) halfway between **Regulus** and the southwestern horizon.

Boötes and **Arcturus** are prominent in the eastern sky, together with the circlet of **Corona Borealis**, framed by Boötes and the neighbouring constellation of **Hercules.** Between Leo and Boötes lies **Coma Berenices**, notable for being the location of the open cluster **Melotte 111** and the **Coma Cluster** of galaxies (Abell 1656). There are about 1,000 galaxies in this cluster, which is located near the North Galactic Pole, where we are looking out of the plane of the Galaxy and are thus able to see deep into space. Only about ten of the galaxies in the Coma Cluster are visible with the largest amateur telescopes.

Farther south, **Crux** is now nearly upright, with the two brightest stars of **Centaurus** (α and β Centauri, **Rigil Kentaurus** and **Hadar**, respectively) to its southeast. A third star in the system, known as **Proxima Centauri**, is much fainter, and is actually the closest star to the Solar System at a distance of about 4.3 light-years. In recent years, it has been found to host an exoplanet, which orbits Proxima every 11.2 days. The magnificent globular cluster, **Omega Centauri**, is high in the sky.

West of Crux, both the **Southern Pleiades** and the **Eta Carinae Nebula** are clearly seen. Farther west, the **False Cross** is beginning to decline towards the horizon. **Canis Major** and brilliant **Sirius** are also descending in the west. **Achernar** (α Eridani) is skimming the southern horizon for observers at 30°S, but **Peacock** (α Pavonis) is now slightly higher and more easily visible. Although the **LMC** is roughly as high as the South Celestial Pole, the **SMC** and **47 Tucanae** are rather low in the south. In the east, the whole of **Scorpius** is clear of the horizon with the neighbouring zodiacal constellation **Libra** preceding it along the ecliptic. The dense regions of the Milky Way in **Sagittarius** become visible later in the night.

A

May

May – Introduction

Meteors

The *η-Aquariids* are one of the two meteor showers associated with Comet 1P/Halley, Halley's Comet, last seen in 1986 and not due until 2061. The other shower arising from the cometary debris is the Orionids, in October. The η-Aquariids are not particularly favourably placed for observers in the northern hemisphere, because the radiant is near the celestial equator, near the Water Jar in Aquarius, well below the horizon until late in the night (around dawn). However, meteors may still be seen in the eastern sky even when the radiant is below the horizon.

The peak of the shower on 6 May coincides with a bright waning gibbous Moon. The Moon rises just after midnight, observing conditions will be favourable before then although more meteors are likely to be seen in the early hours of the morning. Maximum hourly rate is about 40 per hour (although around half that number will be seen from the northern hemisphere) and a large proportion (about 25 per cent) of the meteors leave persistent trains which can last for several seconds to minutes.

In 2024, the birth of three galaxies present in the early Universe, 13.3–13.4 billion years ago, was observed with NASA's **James Webb Space Telescope (JWST)**. Researchers at the University of Copenhagen have captured the accumulation of large clouds of gas accreting onto a baby galaxy. The JWST imaged the infrared light emitted by the cool neutral hydrogen gas, a major ingredient of star formation.

The planets

Mercury moves from *Pisces* into *Aries* and eventually *Taurus*, brightening from mag. –0.8 to –2.4 and then dimming to –0.6. In the latter half of the month Mercury appears shortly after sunset. *Venus* starts the month in Taurus and moves into *Gemini*, visible after sunset (mag. –3.9 to –4.0). *Mars* starts in Pisces and passes into Aries, appearing at dawn a few hours before sunrise (mag. 1.2 to 1.3). *Jupiter* lies in Gemini and is above the horizon from sunset until around midnight (mag. –2.0 to –1.9). *Saturn* lies on the border between *Cetus* and Pisces, visible just before sunrise (mag. 0.9). *Uranus* in Taurus approaches the Sun and moves into the dawn sky after 22 May (mag. 5.8). *Neptune* is low in the east before sunrise in Pisces at mag 7.8.

M

Sunrise and sunset

City	Date	Sunrise	Sunset
Buenos Aires, Argentina			
	May 01	10:30	21:12
	May 31	10:52	20:51
Cape Town, South Africa			
	May 01	05:21	16:06
	May 31	05:43	15:46
London, UK			
	May 01	04:33	19:24
	May 31	03:50	20:07
Los Angeles, USA			
	May 01	13:04	02:37
	May 31	12:43	02:59
Nairobi, Kenya			
	May 01	03:28	15:32
	May 31	03:29	15:32
Sydney, Australia			
	May 01	20:29	07:15
	May 31	20:51	06:55
Tokyo, Japan			
	May 01	19:49	09:27
	May 31	19:27	09:51
Washington, DC, USA			
	May 01	10:10	00:01
	May 31	09:45	00:27
Wellington, New Zealand			
	May 01	19:08	05:28
	May 31	19:36	05:01

NB: the times given are in Universal Time (UT)

The Moon's phases and ages

Northern Hemisphere

Full Moon

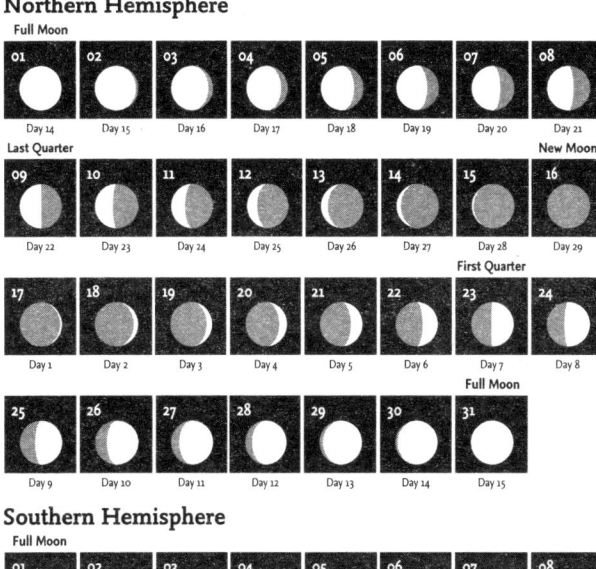

Southern Hemisphere

Full Moon

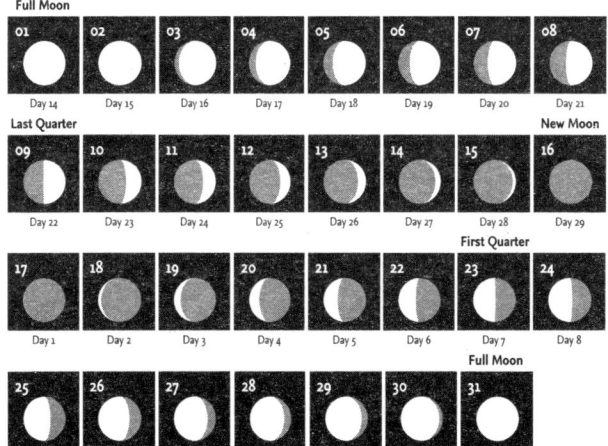

The Moon

The Moon in May

The Moon is full on the first day of the month. **Antares** is 0.5°N of the waning gibbous Moon on 4 May. The Last Quarter Moon will be visible on 9 May and 16 May brings the New Moon. On 19 May **Venus** (mag. −3.9) lies 2.9°S of the waxing crescent Moon. A day later the Moon swings 3.1°N of **Jupiter** (mag. −1.9) and 3.4°S of **Pollux**. On 23 May **Regulus** is less than a tenth of a degree apart from the First Quarter Moon and on 27 May **Spica** is 1.9°N of the Moon. The end of the month brings a Full Moon with **Antares** 0.4°N.

AI in Astronomy

AI machine learning tools were used to find 69 exoplanets, with results published in May 2023. Typically observations of exoplanets are confirmed by further observations using a different detection method. Current methods of finding exoplanets include looking for dips in the host star's brightness as planets pass between Earth and the star, and measuring periodic wobbles in stars due to gravitational perturbations from the orbiting planets. The accuracy of machine learning techniques has improved to the extent that they can be used to validate an observation, improving the efficiency of exoplanet detection.

An AI algorithm called the HelioLinc3D program was used to find a Potentially Hazardous Asteroid (PHA) in August 2023. The asteroid, a 180-m wide rock called **2022 SF289**, is predicted to pass the Earth within a distance of 225,000 km (closer than the Moon). The AI program will be used at the Vera C. Rubin Observatory, currently being built on a mountaintop in Chile, to find more PHAs.

An international collaborative research team are also training a machine learning model to look for biosignatures in rock and crystals in a salt flat in the Atacama Desert, Chile. Their system can detect biosignatures up to 88 per cent of the time. This model will accelerate the search for life on **Mars**, by guiding rovers such as NASA's **Perseverance** rover currently investigating Jezero Crater – a site thought to have once harboured water.

Jezero Crater on Mars, captured by NASA's Perseverance rover, 2021.

M

Calendar for May

01	17:23	Full Moon
04	02:20	Antares 0.5°N of the Moon. An occultation will be visible from Antarctica, Argentina, Chile and Bolivia.
04	22:30	Moon at apogee = 405,843 km
06		Eta-Aquariid meteor shower maximum
09	21:10	Last Quarter Moon
16	20:01	New Moon
17	13:48	Moon at perigee = 358,074 km
19	01:50	Venus (mag. −3.9) 2.9°S of the Moon
20	12:39	Jupiter (mag. −1.9) 3.1°S of the Moon
20	16:30	Pollux 3.4°N of the Moon
23	06:41	Regulus 0.0°N of the Moon
23	11:11	First Quarter Moon
27	14:09	Spica 1.9°N of the Moon
31	08:32	Antares 0.4°N of the Moon. An occultation will be visible from Argentina, eastern Australia, Chile and New Zealand.
31	08:45	Full Moon

6 May: *The waning gibbous Moon in Sagittarius in the south, the Eta Aquariids meteor shower in the southeast in the dawn sky (as seen from London).*

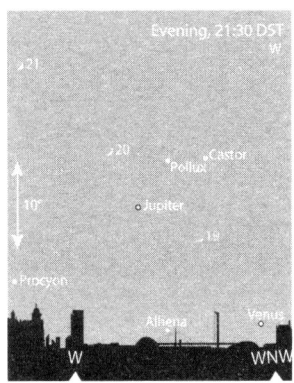

19–21 May: *The waxing crescent Moon in Gemini with Jupiter and Venus, Pollux and Castor. Procyon close by. The Moon moves into Cancer (as seen from central USA).*

M

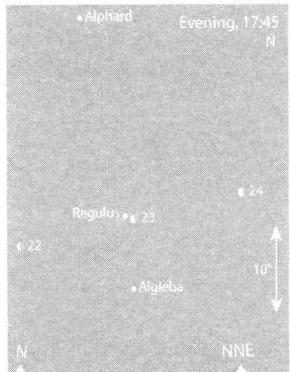

23 May: *The First Quarter Moon close to Regulus in Leo (as seen from Sydney).*

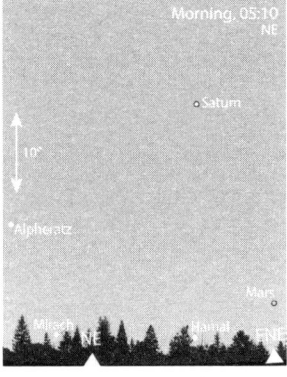

30 May: *Saturn rising in the east along with Mars just before sunrise. Alpheratz (α And) sits lower down (as seen from Sydney).*

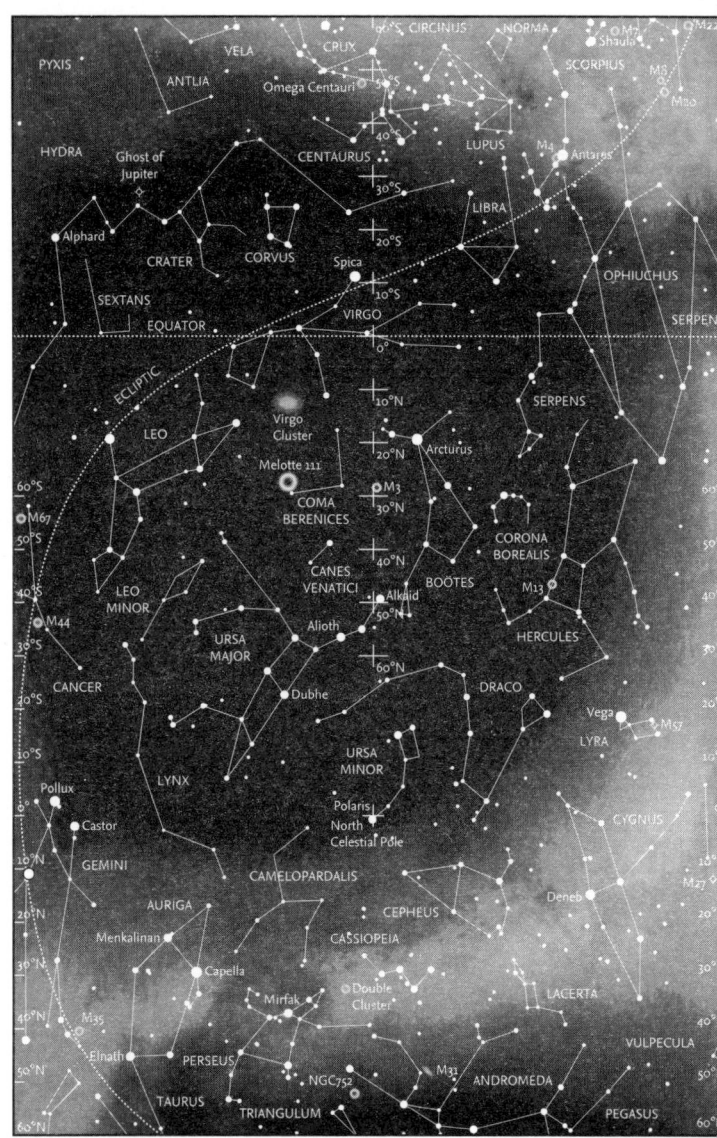

May – Looking North

Cassiopeia is now low over the northern horizon and, to its west, the southern portions of both *Perseus* and *Auriga* are becoming difficult to observe, although the *Double Cluster* between Perseus and Cassiopeia is still clearly visible. Each cluster is composed of more than 300 supergiants many times hotter than the Sun. Some of these stars are nearing the end of their lives and have evolved into red supergiants. The constellations of *Lyra*, *Cepheus*, *Ursa Minor* and the whole of *Draco* are well placed in the sky. *Gemini*, with *Castor* and *Pollux*, is sinking towards the western horizon. *Capella* (α Aurigae) and the three stars in the asterism the Kids are still clear of the horizon.

In the east, two of the stars of the Summer Triangle, *Vega* (α Lyrae) and *Deneb* (α Cygni), are clearly visible, and the third star, *Altair* in *Aquila*, is beginning to climb above the horizon. The whole of *Cygnus* is now visible. The sprawling constellation of *Hercules* is high in the east and the brightest globular cluster in the northern hemisphere, M13, is visible to the naked eye on the western side of the asterism known as the Keystone.

Three faint constellations may be identified before the lighter nights of summer make them difficult objects. Below Cepheus in the northeastern sky is the zig-zag constellation of *Lacerta*, while to the west, above Perseus and Auriga is *Camelopardalis* and, farther west, the line of faint stars forming *Lynx*.

Later in the night (and in the month) the westernmost stars of *Pegasus* begin to come into view. High overhead, *Alkaid* (η Ursae Majoris), the last star in the 'tail' of the Great Bear, is close to the zenith for observers at 50°N, while the main body of the constellation has swung round into the western sky (the 'legs' and 'paws' of the 'bear').

M

May – Looking South

Early in the night, the constellation of **Virgo**, with **Spica** (α Virginis), lies due south, with **Leo** and both **Regulus** and **Denebola** (α and β Leonis, respectively) to its west still well clear of the horizon. Virgo contains the nearest large cluster of galaxies, which is the centre of the Local Supercluster, of which the Milky Way galaxy forms part. The Virgo Cluster contains some 2,000 galaxies, the brightest of which are visible in amateur telescopes. Later in the night, the rather faint zodiacal constellation of **Libra** becomes visible and, to its east, the red-orange star **Antares** (α Scorpii) begins to climb up over the horizon. (As **Orion** sinks in the west, so **Scorpius** rises in the southeast. This recalls one of the many legends about Orion, in which he is pursued by a scorpion, sent to distract him while fighting.)

Arcturus in **Boötes** is high in the south for northern observers, with the distinctive circlet of **Corona Borealis** clearly visible to its east. The brightest star (α Coronae Borealis) is known as **Alphecca**. The large constellation of **Ophiuchus** (which actually crosses the ecliptic and is thus the 'thirteenth' zodiacal constellation) is climbing into the eastern sky. Before the constellation boundaries were formally adopted by the International Astronomical Union in 1930, the southern region of Ophiuchus was regarded as forming part of the constellation of Scorpius, which had been part of the zodiac since antiquity. It lies between the two portions of **Serpens** (the only constellation to be divided into two parts). **Serpens Caput** (the Head of the Serpent) lies west of Ophiuchus, between it and Boötes, and **Serpens Cauda** (the Tail of the Serpent) is to the east, between Ophiuchus and the Milky Way.

Farther south, between Scorpius and the two brightest stars in **Centaurus** (**Rigil Kentaurus** and **Hadar**), lies the constellation of **Lupus**, just east of the meridian and lying along the Milky Way. The whole of **Sagittarius** is now clearly seen in the east at lower latitudes, with **Corona Australis** beneath it and Scorpius higher above. **Crux** is now more-or-less 'upright' and visible for viewers at latitudes below 20°N, with the small constellation of **Musca** below it. Below the bright pair of stars in Centaurus (α and β Centauri) is the constellation of **Triangulum Australe**, visible at latitudes below 25°N and a much larger and more

M

striking constellation than its counterpart (Triangulum) in the north. Lying between Rigil Kentaurus and Triangulum Australe is the very tiny and indistinct constellation of **Circinus**.

The *Eta Carinae Nebula*, the *Southern Pleiades* and the *False Cross* lie in the west. The stars of *Vela* and *Carina* are becoming lower in the southwest, following the constellation of *Puppis* and brilliant *Canopus* (α Carinae) down towards the horizon. Puppis and the *Large Magellanic Cloud* (LMC) are close to the horizon for anyone at the latitude of 30°S (about the latitude of Sydney in Australia), where *Achernar* (α Eridani) is now too low to be seen. The *Small Magellanic Cloud* (SMC) is beginning to rise higher in the sky. There are several, small, relatively faint constellations in this part of the sky. The most distinct is probably *Pavo*, with its single bright star (α Pavonis), known as *Peacock*. Other constellations are *Apus*, *Chamaeleon*, *Octans* (which actually includes the South Celestial Pole), *Mensa* and *Volans*.

The highest observatory in the world

The University of Tokyo Atacama Observatory (TAO), located on the summit of Cerro Chajnantor in the Atacama Desert in Chile, is the highest observatory in the world, at an altitude of 5,640 m. The Earth's atmosphere impacts observations, turbulence affects the quality of images – astronomers refer to this effect as 'seeing'. Many large terrestrial telescopes employ adaptive optics to mathematically subtract this unwanted noise and improve the clarity of images. The best images can be captured from space-based telescopes; however, launching telescopes into space vastly increases the cost and limits the size of the mirror used to collect light, called the primary objective.

The TAO took 26 years to plan and construct, and has been operational since May 2024, working in the infrared region of the spectrum. At its altitude humidity is vastly reduced, a significant benefit as water vapour affects infrared observations. The telescope will be used to observe dusty circumstellar environments, including embryonic planetary systems. However, the risk of altitude sickness is high for astronomers working there, and a period of acclimatization may be required before the work of probing exoplanet systems can begin.

The University of Tokyo Atacama Observatory (TAO).

Cerro Chajnantor in the Atacama Desert, Chile – the site of the TAO.

June

June – Introduction

Around the summer solstice (21 June) in the northern hemisphere (winter solstice in the southern hemisphere), a form of twilight persists throughout the night, due to the Sun not setting as far below the horizon. At high latitudes the sky remains so light that most of the fainter stars and constellations are invisible. But there is one compensation during these light nights: even southern observers may be lucky enough to witness a display of noctilucent ('night-shining') clouds (NLC). These are highly distinctive clouds shining with an electric-blue tint, observed in the sky in the direction of the North Pole. They are the highest clouds in the atmosphere, occurring at altitudes of 80–85 km, and they are composed of ice crystals. They are only visible during summer nights shortly after sunset and before sunrise, for up to six weeks on either side of the solstice, but the clouds themselves remain illuminated by sunlight. They act like high mirrors, reflecting sun rays when the Sun is below the horizon.

There is a commonly held view that the weather in June is hot because the Earth is then closest to the Sun, but this is completely wrong. The Earth's orbit is not a true circle, but an ellipse. In 2026 the distance varies from 147,099,900 km (0.9833 AU) on 3 January (at perihelion) to 152,087,778 km (1.01664 AU) on 6 July (at aphelion). So the Earth is actually closest to the Sun in January. The winter solstice (for northern-hemisphere residents) when the Sun is lowest in the sky, occurs on 21 December in 2026. At the summer solstice (winter for those in the southern hemisphere) on 21 June, it is actually farther from the Sun, reaching its farthest point on 6 July.

The seasons, with warm summers and cold winters, are solely caused by the tilt of the Earth's axis to the plane of its orbit around the Sun – the ecliptic. As such, the northern hemisphere is most strongly tilted towards the Sun in June and July, and the southern hemisphere in November and December. At the equinoxes (20 March and 23 September in 2026), the Earth's axis is tilted so that the planet is 'side-on' to the Sun, so day and night are of (approximately) equal length, and the Sun's heat is evenly distributed between the two hemispheres.

In June 2024, the Russian space agency Roscosmos broke a defunct satellite into almost 200 pieces of debris, adding to the ever-growing amount of space junk orbiting the Earth. The six US astronauts onboard the **International Space Station (ISS)** took shelter for about an hour in a spacecraft docked with the ISS in case they needed to depart in an emergency. There are approximately 25,000 pieces of debris larger than 10 cm; collisions with satellites can create an avalanche effect of even more debris called the Kessler effect. The safest option for decommissioning satellites is to boost them up into a high graveyard orbit 36,000 km above the Earth, to avoid collisions with operating satellites lower down.

The planets
Mercury is in *Gemini* and appears after sunset as the month progresses (mag. −0.5 to 2.0). On 15 June it is at eastern elongation (mag. 0.5). On 25 June Mercury (mag. 1.3) passes 3.8°S of *Jupiter* (mag. −1.8). *Venus* starts the month in *Gemini*, 12 days later it moves into *Cancer* and by the end of the month it is in *Leo* and visible in the evening (mag. −4.0 to −4.1). On 7 June Venus is 4.6°S of *Pollux* and on 9 June it sits 1.6°N of Jupiter (mag. −1.9). *Mars* is in *Aries* and progresses into *Taurus* in the last week of the month. It's visible up to a few hours before sunrise (mag. 1.3 to 1.4). On 28 June Mars is 4.3°S of the *Pleiades* (mag. 1.3). Jupiter moves through Gemini and into Cancer, visible from sunset until around 22:00 (mag. −1.9 to −1.8). *Saturn* sits in *Pisces*, appearing a few hours after midnight (mag. 0.9 to 0.8). *Uranus* is in *Taurus* and only observable around an hour before sunrise (mag. 5.8). *Neptune* lies in Pisces and above the horizon several hours after midnight (mag. 7.8).

J

Sunrise and sunset

City	Date	Sunrise	Sunset
Buenos Aires, Argentina			
	Jun. 01	10:52	20:51
	Jun. 30	11:02	20:54
Cape Town, South Africa			
	Jun. 01	05:43	15:45
	Jun. 30	05:53	15:48
London, UK			
	Jun. 01	03:49	20:08
	Jun. 30	03:47	20:21
Los Angeles, USA			
	Jun. 01	12:43	02:59
	Jun. 30	12:45	03:08
Nairobi, Kenya			
	Jun. 01	03:29	15:32
	Jun. 30	03:35	15:38
Sydney, Australia			
	Jun. 01	20:51	06:54
	Jun. 30	21:01	06:57
Tokyo, Japan			
	Jun. 01	19:27	09:51
	Jun. 30	19:28	10:01
Washington, DC, USA			
	Jun. 01	09:45	00:28
	Jun. 30	09:46	00:37
Wellington, New Zealand			
	Jun. 01	19:37	05:00
	Jun. 30	19:48	05:01

NB: the times given are in Universal Time (UT)

The Moon's phases and ages

Northern Hemisphere

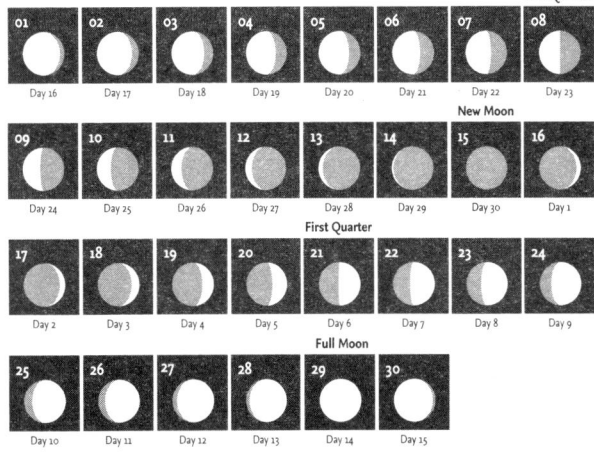

Southern Hemisphere

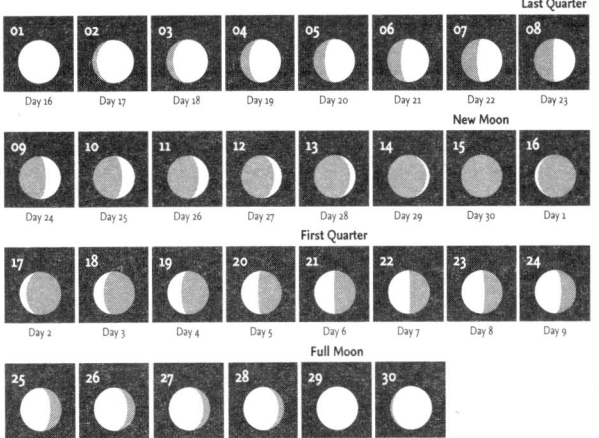

133

The Moon

The Moon in June

The Moon reaches Last Quarter on 8 June. On 13 June the waning crescent Moon passes 1.0°N of the **Pleiades**. A New Moon occurs on 15 June and a day later **Mercury** (mag. 0.5) lies 2.6°S of a thin sliver of the Moon. The following day the waxing crescent Moon is 3.6°S of **Pollux**, 2.5°N of **Jupiter** (mag. −1.8) and 0.3°N of **Venus** (mag. −4.0). On 19 June **Regulus** is 0.3°N of the Moon. The First Quarter Moon is visible on 21 June at the summer solstice. Two days later the waxing gibbous Moon passes 2.2°S of **Spica**. On 27 June **Antares** lies 0.5°N of the Moon and on 29 June the Moon is Full.

Valentina Tereshkova: the first woman in space

Soviet cosmonaut Valentina Tereshkova became the first woman in space on 16 June 1963. She completed 48 orbits of the Earth in her spacecraft **Vostok 6**, spending almost three days in space. Aged 26 at the time, she flew solo and remains the youngest woman to have flown in space. Before training as a cosmonaut, she worked at a tyre factory and textile mill, and took up parachuting in 1959, aged 22. Her chance to train as a cosmonaut was a result of the space race between the Soviet Union and the USA. The director of cosmonaut training, Nikolai Kamanin, was eager to beat the Americans who were screening potential female astronauts in an unofficial program using the same tests as their male counterparts in Project Mercury. The 13 women who passed the selection process later became known as the Mercury 13. While NASA did not allow the women into their astronaut programme, the Russian space agency followed through with their plan.

Five female Soviet cosmonauts began training in early 1963, including Tereshkova. She was nominated to pilot *Vostok 6*. As a tribute to her drive and determination, Kamanin dubbed her 'Gagarin in a skirt'. During her two days, 22 hours and 50 minutes in space, Tereshkova took photographs of the horizon, which were used to identify aerosol layers in the atmosphere. The Russian space agency gained valuable data about the effects of space travel on women. She was awarded the Hero of the Soviet Union medal for her successful mission.

The Vostok 6 *capsule on display at the Science Museum, London in 2016.*

Tereshkova married her colleague, *Vostok 3* cosmonaut Andriyan Nikolayev, in November 1963 and she gave birth to a daughter in June 1964, a little girl whose parents had both been to space. Kamanin had planned for two more women to travel to space, but these plans were cancelled in 1965. The next woman to go into space was Svetlana Savitskaya in 1982, 19 years after Tereshkova.

In June 1983, Sally Ride became the first American astronaut to go into space via the Space Shuttle *Challenger*. Three years later, the same spacecraft was in the news for more tragic reasons; in 1986 it exploded shortly after launch, killing all seven crew members onboard.

J

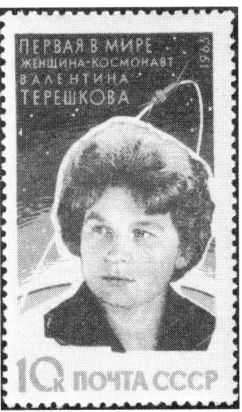

Valentina Tereshkova on a Soviet postage stamp in 1963.

Calendar for June

01	04:32	Moon at apogee = 406,369 km
07	16:17	Venus (mag. −4.0) 4.6°S of Pollux
08	10:00	Last Quarter Moon
09	20:03	Venus (mag. −4.0) 1.6°N of Jupiter (mag. −1.9)
13	13:15	Pleiades 1.0°S of the Moon
14	23:18	Moon at perigee = 357,196 km
15	02:54	New Moon
15	20:00	Mercury at eastern elongation (24.5°E, mag. 0.5)
16	19:32	Mercury (mag. 0.5) 2.6°S of the Moon
17	02:08	Pollux 3.6°N of the Moon
17	06:54	Jupiter (mag. −1.8) 2.5°S of the Moon
17	20:21	Venus (mag. −4.0) 0.3°S of the Moon. An occultation will be visible from Canada, Brazil and Venezuela.
19	14:31	Regulus 0.3°N of the Moon. An occultation will be visible from South Africa, Mozambique, Madagascar and Zimbabwe.
21	08:25	Summer Solstice
21	21:55	First Quarter Moon
23	20:11	Spica 2.2°N of the Moon
25	12:00	Mercury (mag. 1.3) 3.8°S of Jupiter (mag. −1.8)
27	14:32	Antares 0.5°N of the Moon. An occultation will be visible from Antarctica, southeastern Australia, New Zealand and Tasmania.
28	07:11	Moon at apogee = 406,267 km
28	18:32	Mars (mag. 1.3) 4.3°S of the Pleiades
29	23:57	Full Moon

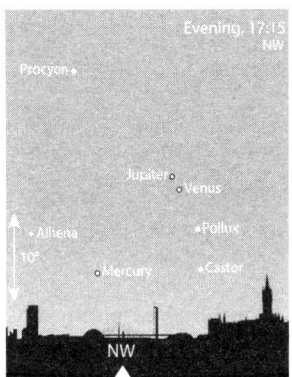

9 June: Venus and Jupiter appear close together in Gemini, with Mercury nearby. Pollux and Castor shine brightly (as seen from Sydney).

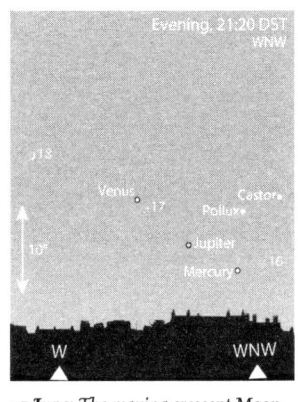

17 June: The waxing crescent Moon passes by Venus in Cancer, Jupiter lies towards the northwest (as seen from London).

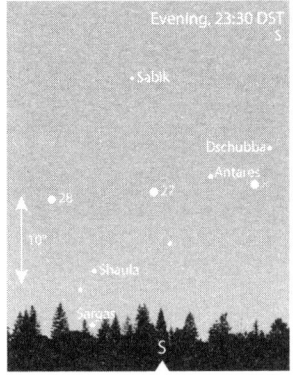

27 June: The waxing gibbous Moon lies to the east of Antares, the rival of Mars (as seen from central USA).

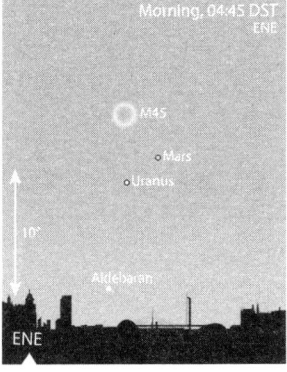

29 June: Mars, Uranus and the Pleiades (M45) all in Taurus rising in the east just before sunrise (as seen from central USA).

J

137

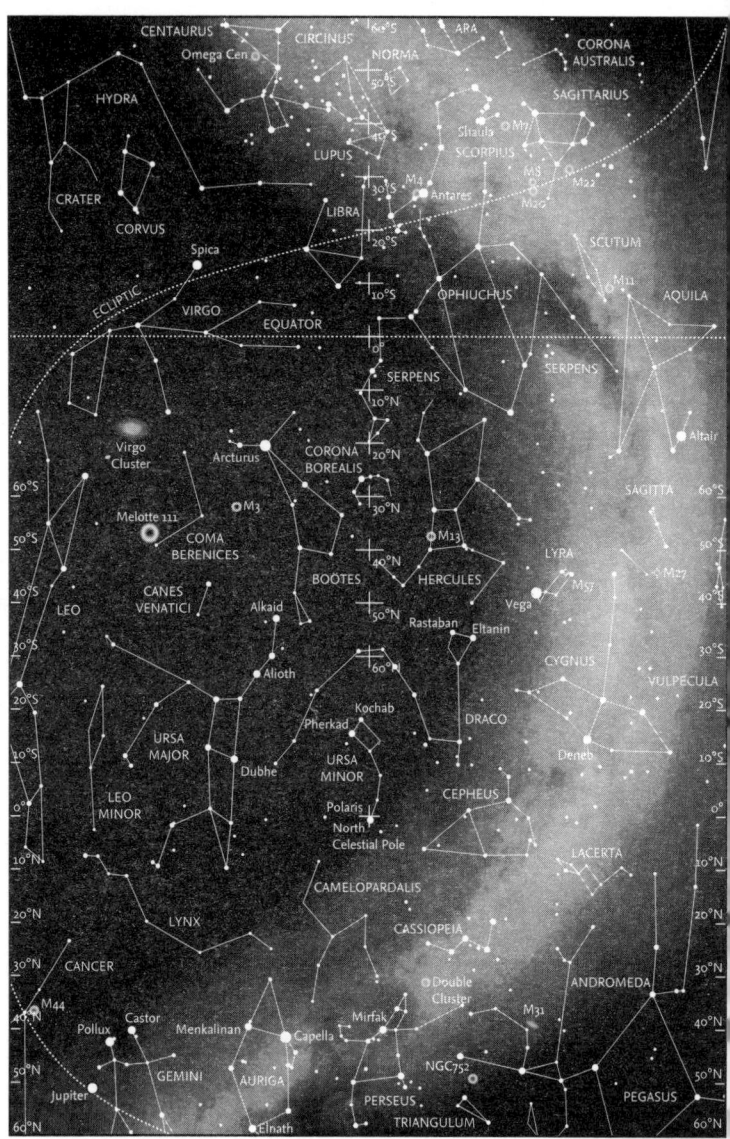

June – Looking North

For those in the northern hemisphere the time around the summer solstice (21 June) is frustrating for observing, because a form of twilight persists throughout the night. The sky remains so light that most faint stars and constellations are invisible unless conditions, such as the lack of light pollution, are particularly favourable. Even the highly distinctive set of seven stars forming the asterism of the Plough (or Big Dipper) in **Ursa Major**, now in the western sky, may be difficult to see. The constellation of **Boötes** with bright **Arcturus** (α Boötis) is high overhead for northern observers, and best seen when facing south. The sprawling constellation of **Hercules** lies between **Lyra** and Boötes.

The four stars forming the 'head' of **Draco** lie east of the meridian, although only the brightest, **Eltanin** (γ Draconis) and possibly **Rastaban** (β Draconis), are clearly visible. Farther north, in **Ursa Minor**, are **Polaris** itself and the two Guards, **Kochab** (β UMi) and **Pherkad** (γ UMi).

For observers around 50°N, the southern portions of the constellation of **Auriga** are partially lost on the northern horizon, although bright **Capella** (α Aurigae) should still be visible. Much of the neighbouring, fainter constellation of **Perseus** is difficult to make out. The two main stars of **Gemini**, **Castor** (α Geminorum) and **Pollux** (β Geminorum), the star closest to the ecliptic, and occasionally occulted by the Moon, are low on the northwestern horizon. Somewhat higher in the sky and northeast of the meridian is **Cassiopeia**, with **Cepheus** above it.

Two of the stars forming the angles of the distinctive Summer Triangle, **Deneb** (α Cygni) in **Cygnus** and **Vega** (α Lyrae) in Lyra, are clearly visible as, for much of the night, is the third star, **Altair** (α Aquilae) in **Aquila** over in the east. Deneb lies in the Milky Way, at the beginning of the **Great Dark Rift** that runs down the constellation and where obscuring dust prevents us from seeing the dense star clouds of the Milky Way itself. All three stars are larger than the Sun, Deneb being 200 times greater in size than the Sun and up to 200,000 times more luminous. Lying 100 times further away than the other two stars, we see all three having similar apparent brightness in the summer sky. If Deneb was the same distance as Vega and Altair, it would be so bright in our sky it would cast shadows, like the Full Moon.

J

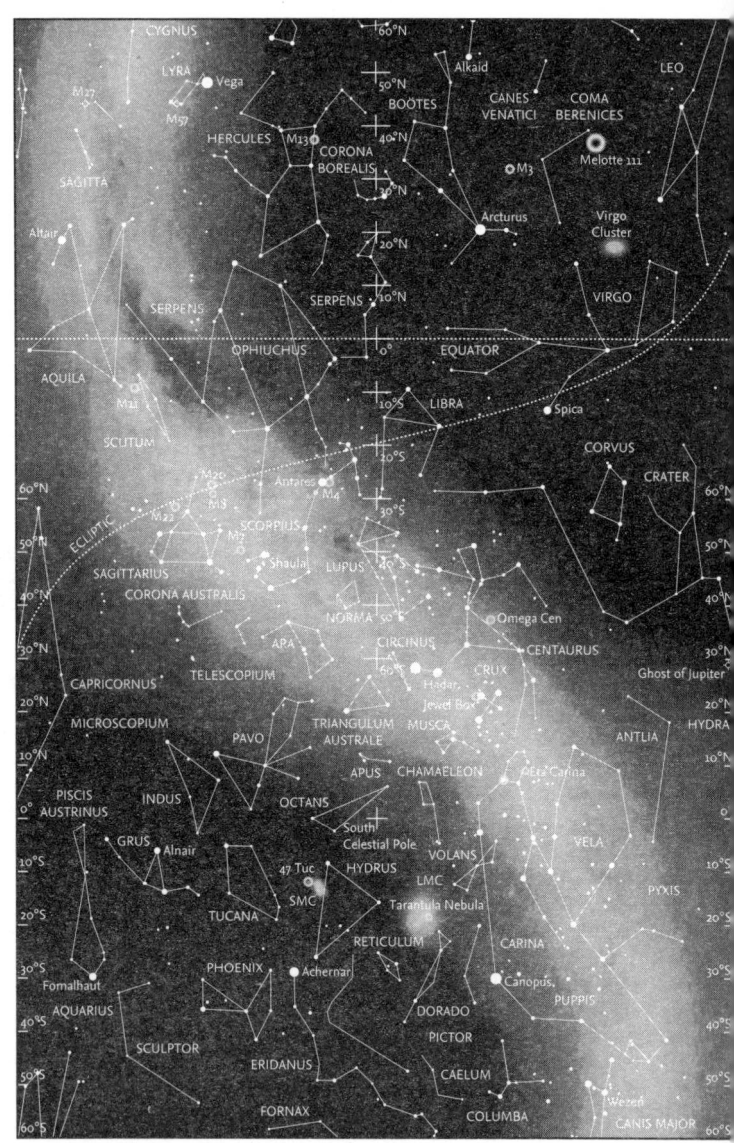

June – Looking South

Despite the persistent twilight, there is still plenty to see after dark. The zodiacal constellation of **Libra** lies almost due south. The fiery red supergiant star **Antares** – the name means the 'rival of Mars' – in **Scorpius** is visible slightly to the east of Libra, but the 'tail' or 'sting' remains below the horizon for northern observers. Higher in the sky is the large constellation of **Ophiuchus** (the Serpent Bearer), lying between the two halves of the constellation of **Serpens**: **Serpens Caput** (Head of the Serpent) to the west and **Serpens Cauda** (Tail of the Serpent) to the east. The ecliptic runs across Ophiuchus, and the Sun actually spends far more time in the constellation than it does in the 'classical' zodiacal constellation of Scorpius, a small area of which lies between Libra and Ophiuchus. The constellation of **Virgo** (and bright **Spica**) is now well to the southwest after midnight, along the ecliptic.

At this time of the year, the next constellation in the zodiac, **Sagittarius**, is becoming more visible, having risen in the east. The main body of the constellation forms the asterism known as the Teapot, but the constellation also has a long, curving chain of faint stars to the south, rather similar to the long line of stars below Scorpius, although curving in the opposite direction. This chain of stars partially encloses the small constellation of **Corona Australis**. Following **Sagittarius** into the sky is **Capricornus**. For observers in the southern hemisphere, the constellation of **Grus** has now risen above the horizon with the faint constellation of **Indus** between it and **Pavo**. To the east of Grus is **Piscis Austrinus**, although brilliant **Fomalhaut** (α Piscis Austrini) is only just clear of the horizon and becomes clearly visible only later in the night and later in the month.

Higher in the southern sky, the three constellations of **Boötes**, **Corona Borealis** and **Hercules** are now better placed for observation than at any other time of the year. This is an ideal time to observe the fine globular cluster of **M13** in Hercules with binoculars or a telescope.

For observers at the latitude of Sydney in Australia, both **Canopus** (α Carinae) and **Achernar** (α Eridani) are skimming the southern horizon. Although the whole of **Carina** is visible, all of **Eridanus** (except Achernar) is hidden below the horizon. The **Large Magellanic Cloud** (LMC) is low, although the

Small Magellanic Cloud (SMC), the globular cluster *47 Tucanae* and *Hydrus* are now rather higher. *Alphard* (α Hydrae) is on the horizon and the remainder of the long constellation of *Hydra* is clearly seen, as are the two constellations of *Crater* and *Corvus* to its north. The *False Cross* between Carina and *Vela* is beginning to descend in the west, but Vela itself is clearly visible. The whole of both *Crux* and *Centaurus* are clearly seen, as are the magnificent globular cluster, *Omega Centauri*, and the constellation of *Lupus*, closer to the zenith. The constellation of *Triangulum Australe* is on the meridian, roughly halfway between the South Celestial Pole and the zenith.

Mission to Uranus

In May 2024, NASA held a meeting to discuss the **Uranus Orbiter Probe**, a mission to the icy giant Uranus. The mission had been given priority status in 2022, with the key goals of analysing the upper atmosphere, the magnetic field and the effects of the extreme 98° tilt of the planet. The probe will be equipped to beam back data of some of the moons, contributing to the search for watery worlds beyond Saturn. Ice giants exhibit characteristics that overlap those of terrestrial planets and gas giants, and scientists have yet to learn about the evolution of the internal structures of Uranus and Neptune and their place in the story of the Solar System. The last visitor to Uranus was *Voyager 2*, which flew past in January 1986, taking more than 7,000 photographs. The probe revealed 11 new moons, two new rings around the planet, and a displaced and highly tilted magnetic field.

NASA are keen to develop and launch the probe within the next decade. Scientists involved in the project are calling for a collaboration with the European Space Agency (ESA) to meet their target deadline; funding for the project is due to start by 2027. If successful, the probe will take 12–15 years to reach Uranus. In 2004, ESA and NASA partnered for their **Cassini-Huygens** mission to Saturn, a hugely successful mission that lasted almost 20 years following its launch in October 1997.

The five largest moons of Uranus: Miranda, Ariel, Umbriel, Titania and Oberon, imaged by NASA's Voyager 2, first imaged in 1986.

J

July

July – Introduction

Northern observers will continue to experience light nights during July, although there is still the chance that they may be able to see noctilucent clouds during the first half of the month. The noctilucent cloud 'season' in the north lasts about 6–8 weeks, centred on the summer solstice.

The Earth reaches aphelion, the furthest point from the Sun in its yearly orbit, at 17:30 UT on 6 July. Its distance from the Sun is then 1.01664 AU (which equals 152,087,778 km).

Meteors

July brings three meteor showers. There are two minor radiants active in the constellations of **Capricornus** and **Aquarius**. Because of their location, however, observing conditions are not particularly favourable for northern-hemisphere observers. The first shower, the α-**Capricornids**, active from 3 July to 15 August (peaking 30 July), has a maximum rate of about 5 per hour; however, it does often produce very bright fireballs. The parent body is Comet 169P/NEAT. The most prominent shower is probably that of the **Southern δ-Aquariids**, which are active from around 12 July to 23 August, with a peak on 30 July, although even then the rate is unlikely to reach 25 meteors per hour. In this case, the parent body may be Comet 96P/Machholz. This year, both shower maxima occur when the Moon is a bright waning gibbous; moonlight will interfere with observations. The **Piscis Austrinids** begin on 15 July and continue until 10 August. Maximum is on 28 July, but the rate is only about 5 per hour. The **Perseids** begin on 17 July and peak on 13 August.

The planets

Mercury is placed in **Gemini** and eventually appears just after sunset until the middle of the month, after which it leads the Sun at sunrise (mag. 2.2 to 5.6 then brightening again to 0.5). **Venus** stays in **Leo** for the month, appearing after sunset (mag. −4.1 to −4.3). On 9 July it passes 0.9°N of **Regulus**, visible until a few hours before midnight. **Mars** sits in **Taurus**, appearing a few hours before sunrise (mag. 1.4 to 1.3). On 4 July **Mars** is 0.1°S of **Uranus** (mag. 5.8). **Jupiter** lies in **Cancer**, visible from sunset until around 20:00. The gas giant approaches the Sun and leads the Sun in the dawn sky after 29 July (mag. −1.8). *Saturn* continues in *Pisces*, climbing above the eastern horizon after midnight (mag. 0.8 to 0.6). Saturn enters retrograde motion on 26 July at mag. 0.7. Uranus in Taurus is lost in the glare of the Sun at sunrise (mag. 5.8). *Neptune* in Pisces appears after midnight and enters retrograde motion on 7 July (mag. 7.8 to 7.7).

Four volunteer crew members emerged from a simulated Mars environment created by NASA on 6 July 2024 after 378 days isolated from the rest of the world. Anca Selariu, Kelly Haston, Ross Brockwell and Nathan Jones were locked inside **Mars Dune Alpha** at NASA's Johnson Space Center in Houston, Texas. The crew were exposed to simulated emergencies, high workload, equipment failures and other stressors a real crew might face on Mars. It is projected that humans may land on Mars as early as the next decade.

J

Sunrise and sunset

City	Date	Sunrise	Sunset
Buenos Aires, Argentina			
	Jul. 01	11:02	20:54
	Jul. 31	10:48	21:12
Cape Town, South Africa			
	Jul. 01	05:53	15:48
	Jul. 31	05:40	16:06
London, UK			
	Jul. 01	03:48	20:21
	Jul. 31	04:23	19:50
Los Angeles, USA			
	Jul. 01	12:45	03:08
	Jul. 31	13:04	02:54
Nairobi, Kenya			
	Jul. 01	03:35	15:38
	Jul. 31	03:37	15:41
Sydney, Australia			
	Jul. 01	21:01	06:57
	Jul. 31	20:49	07:15
Tokyo, Japan			
	Jul. 01	19:28	10:01
	Jul. 31	19:48	09:47
Washington, DC, USA			
	Jul. 01	09:47	00:37
	Jul. 31	10:08	00:20
Wellington, New Zealand			
	Jul. 01	19:48	05:02
	Jul. 31	19:30	05:25

NB: the times given are in Universal Time (UT)

The Moon's phases and ages

Northern Hemisphere

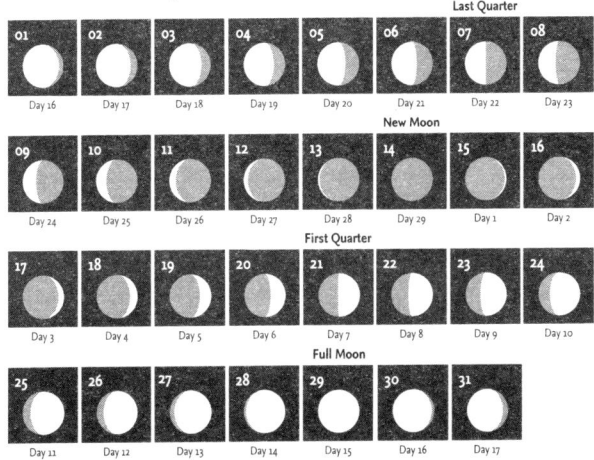

Southern Hemisphere

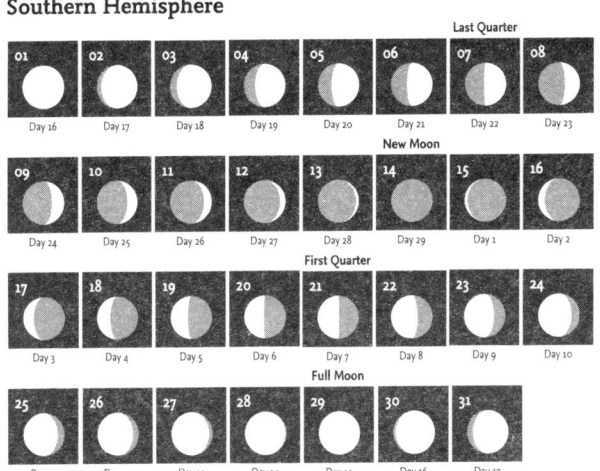

The Moon

The Moon in July

The Moon is Last Quarter on 7 July. On 10 July the waning crescent Moon is 1.1°N of the *Pleiades*. A New Moon takes place on 14 July. Three days later the waxing crescent Moon passes 0.5°S of *Regulus* and 2.0°S of *Venus* (mag. −4.2). On 21 July the First Quarter Moon lies 2.4°S of *Spica*. On 24 July the waxing gibbous Moon approaches *Antares*, sitting 0.6°S of the red supergiant star. A Full Moon dominates the night sky on 29 July in *Capricornus*.

Meteor explodes over Manhattan

Tuesday 16 July 2024 was another ordinary morning in Manhattan, New York, until a loud boom was heard with accompanying reverberations. NASA Meteor Watch identified the cause: a meteor or space rock first seen around 78 km above Upper Bay, then descending at a steep angle, passing over the Statue of Liberty and exploding around 46 km above midtown Manhattan. Disintegrating meteors that are very bright are called fireballs and daylight observations of them

Fireball over the Atacama Large Millimeter/submillimeter Array in the Atacama Desert, Chile, 2014.

are very rare. The American Meteor Society receives hundreds of fireball reports each year via their online form; they advise the public only to report a one-off bright (possibly loud) event that lasts less than 30 seconds and looks like a shooting star.

Henrietta Swan Leavitt: Variable stars

Stars are categorized in a variety of different types; our Sun is described as a yellow dwarf star. Henrietta Swan Leavitt (1868–1921) joined Harvard College Observatory in Massachusetts, USA, as a volunteer in 1893; she later went on to the photometry department. Leavitt investigated variable stars, which are stars that exhibit periodic fluctuations in their brightness. She compared photographic plates of the same region of sky taken days apart at an observatory in Peru.

She discovered an extraordinary 1,777 variable stars in the Large and Small Magellanic Clouds. Of these, 47 were categorized as 'cluster variables' and then later, Cepheid variables. The stars exhibited periodic variability ranging from 1–120 days, and all presented a rapid increase in brightness, followed by a gradual drop. It is now known they are yellow-type stars with pulsations in their outer layers. Leavitt realized that Cepheid stars with longer periods of variability were brighter on average than those with shorter periods. There was a relationship between period and luminosity (their power output). They were all situated at similar distances from Earth due to their location in the Magellanic Clouds, which meant she could exclude distance as a factor for change in brightness. She published her results in 1908, and since then Cepheids have been used as standard candles for estimating distances to galaxies; a crucial rung of the cosmic distance ladder. Edwin Hubble used Leavitt's findings, and the intrinsic brightnesses of the Cepheids determined by Danish astronomer Ejnar Hertzsprung, to determine the relationship between the distance and recessional velocity of distant galaxies, which led to the explosive realization that the Universe was expanding.

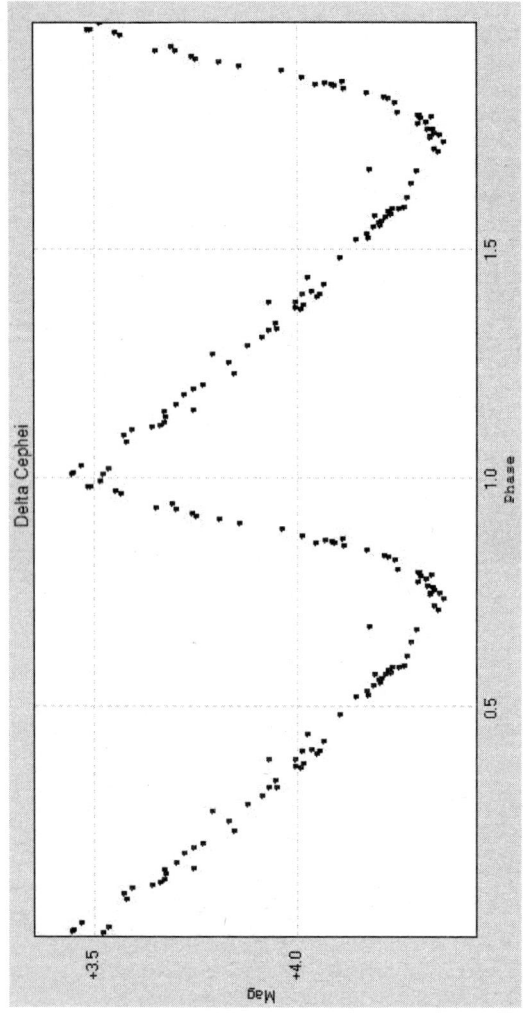

The light curve of the star Delta Cephei, exhibiting regular variability in brightness caused by pulsations in its outer layers.

China's lunar probe **Chang'e-5** has discovered traces of water in the lunar regolith – the Moon's soil. The *Chang'e-5* rover delivered rock and soil samples to Earth in 2020. Scientists have revealed that water is present as hydrated salts in sunlit regions of the Moon; the rover surveyed a young lava plain in the lunar mountains in the northern part of the Ocean of Storms.

A rock has taken geologists by surprise – on **Mars**. NASA's *Perseverance* rover has been exploring Jezero Crater since it landed in February 2021; its mission is to look for possible biosignatures in a region thought to have harboured a lake billions of years ago (*jezero* means 'lake' in some Slavic languages). The *Perseverance* team found a vein-filled rock with unusual markings resembling 'leopard spots' – images were released on 25 July 2024. To fully understand the origins of the spots, the rock would need to be brought back to Earth, a mission currently being planned by the European Space Agency and NASA.

J

Calendar for July

06	17:30	Earth at aphelion (152,087,778 km = 1.01664 AU)
07	19:29	Last Quarter Moon
09	14:36	Venus (mag. −4.1) 0.9°N of Regulus
10	22:54	Pleiades 1.1°S of the Moon
13	07:50	Moon at perigee = 359,111 km
14	09:43	New Moon
17	00:07	Regulus 0.5°N of the Moon
17	16:31	Venus (mag. −4.2) 2.0°N of the Moon
21	03:21	Spica 2.4°N of the Moon
21	11:06	First Quarter Moon
24	21:00	Antares 0.6°N of the Moon
25	16:45	Moon at apogee = 405,549 km
29	14:36	Full Moon
28		Piscis Austrinid meteor shower maximum
30		α-Capricornid meteor shower maximum
30		Southern δ-Aquariid meteor shower maximum

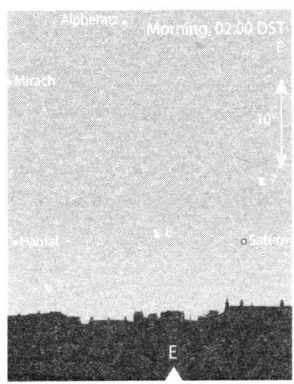

8 July: Last Quarter Moon in Pisces with Saturn, rising in the east after midnight (as seen from central USA).

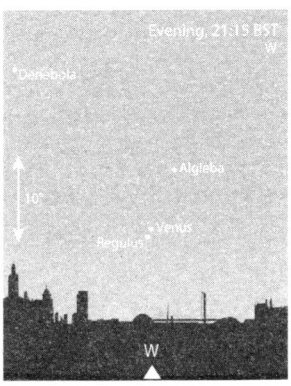

9 July: Venus close to Regulus in Leo, shortly after sunset (as seen from London).

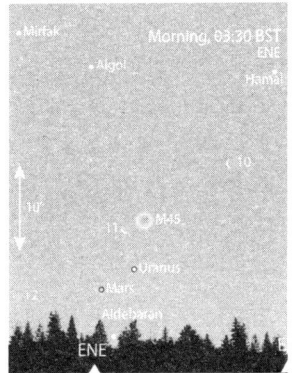

11 July: The waning crescent Moon in Taurus close to the Pleiades (M45), Uranus and Mars shortly before sunrise. Aldebaran and Mirfak (Algenib) are nearby (as seen from London).

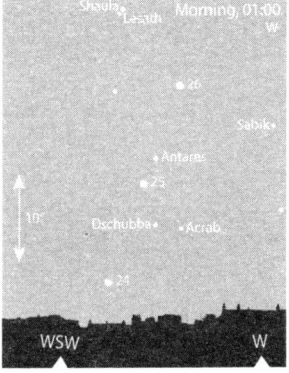

25 July: Waxing gibbous Moon south of Antares. Sabik (η Oph) and Saik (ζ Oph) of Ophiuchus sit westwards (as seen from Sydney).

J

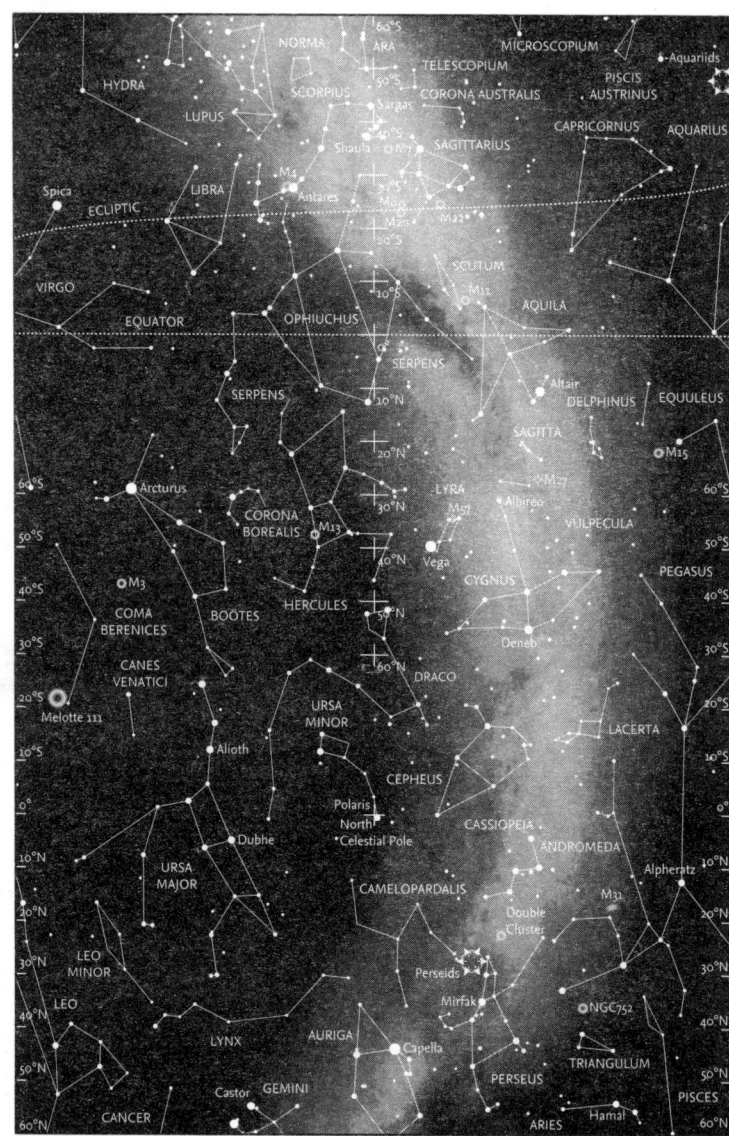

July – Looking North

As in June, light nights and the chance of observing noctilucent clouds persist throughout July, but later in the month (and particularly after midnight) some of the major constellations are more easily seen. The brilliant star **Vega** (α Lyrae) is now shining high overhead and the constellations of **Hercules** and **Lyra** are on opposite sides of the meridian, not far from the zenith for observers at 40°N, while it is the head of **Draco** that is near the zenith for observers at 50°N. In the northwestern sky are the constellations of **Corona Borealis** and **Boötes**, the latter with brilliant, orange-tinted **Arcturus**.

The stars of the Milky Way are now running from north to south on the eastern side of the meridian. The constellation of **Cygnus** is high in the sky. The star **Albireo** (β Cygni) marks the Swan's beak and through a telescope it can be seen to be an optical double: the brighter star is an orange giant 100 times more luminous than the Sun; the second star is blue-white and even more luminous. For observers at the equator, it is the giant constellation of **Ophiuchus** that is at the zenith, with the two parts of **Serpens** (**Serpens Caput** to the west, and **Serpens Cauda** to the east, among the clouds of the Milky Way). Another star in the Summer Triangle with Vega, **Altair** (α Aquilae) in **Aquila**, is similarly high in the sky.

Ursa Major is now clearly visible in the northwest, and on the opposite side of the meridian is the constellation of **Cepheus**, with its base in the Milky Way. **Cassiopeia** lies in the Milky Way on the opposite side of the North Celestial Pole and **Polaris** in **Ursa Minor**. If the sky is dark and clear, you may be able to make out the small, faint constellation of **Lacerta**, lying across the Milky Way between Cassiopeia and Cygnus. In the east, the stars of **Pegasus** are now well clear of the horizon, with the main line of stars forming **Andromeda** roughly parallel to the horizon. **Alpheratz** (α Andromedae) is actually the star at the northeastern corner of the **Great Square of Pegasus**.

The faint constellations of **Camelopardalis** and **Lynx** lie to the west and slightly farther south. **Perseus**, with the famous variable star, **Algol**, is beginning to climb higher in the sky and observers at mid-northern latitudes will find that they can now more clearly see **Capella** (α Aurigae) and the northernmost portion of **Auriga**.

J

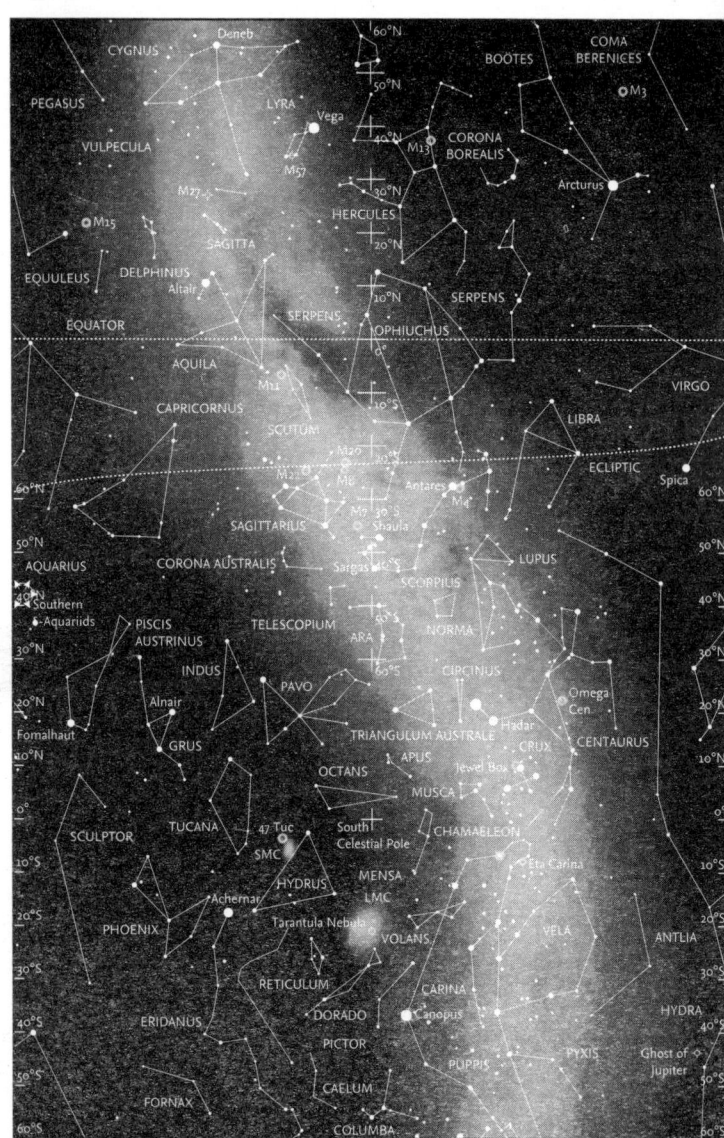

July – Looking South

Although part of the constellation remains hidden, this is perhaps the best time of year to see **Scorpius**, with deep red **Antares** (α Scorpii), glowing just above the southern horizon. **Sagittarius**, with the distinctive asterism of the Teapot, and the dense star clouds of the centre of the Milky Way, are just visible in the south, although along with Scorpius this constellation is low in the sky for northern observers. The **Great Dark Rift** – actually dust clouds that hide the more distant stars – runs down the Milky Way from **Cygnus** towards Sagittarius. Near it are two emission nebulae: **M8** (the Lagoon Nebula) and **M20** (the Trifid Nebula). Towards its northern end is the small constellation of **Sagitta** and the planetary nebula **M27** (the Dumbbell Nebula) in **Vulpecula**. The sprawling constellation of **Ophiuchus** lies close to the meridian for a large part of the month, separating the two halves of the constellation of **Serpens**.

In the east, the bright Summer Triangle, consisting of **Vega** in **Lyra**, **Deneb** in **Cygnus** and **Altair** in **Aquila**, begins to dominate the southern sky.

The constellation of **Hercules** lies on the opposite side of the meridian to Lyra; both are well above the horizon. Farther east, the whole extent of both the zodiacal constellations of **Aquarius** and **Capricornus** is visible. Above Aquila, farther along the Great Rift is the small constellation of **Scutum**, with **M11**, the Wild Duck Cluster. To its east is the tiny but distinctive constellation of **Delphinus**; in the western sky, the constellation of **Boötes** and **Arcturus** (α Boötis) are beginning to approach the horizon, but the circlet of stars that constitute **Corona Borealis** are still clearly seen. Even farther west, the whole of **Virgo** is visible, with **Libra** above it. The stars of **Lupus** and the outlying stars of **Centaurus** (including the great globular cluster known as **Omega Centauri**) lie farther west.

In the southern hemisphere, although the **Large Magellanic Cloud** (LMC) is almost due south, it is very low. Slightly farther west, the **False Cross** is nearing the horizon. **Crux** and the adjoining constellation of **Musca** are now low on the horizon for observers at the equator, and the **Small Magellanic Cloud**

(SMC) in **Hydrus** is actually below it. For observers farther south, **Achernar** (α Eridani) is now clearly visible, as is the constellation of **Phoenix** to the east. Only observers in the extreme south, however, will be able to see the whole of **Vela** and **Carina** as well as **Canopus** (α Carinae). **Grus** and **Piscis Austrinus** (with brilliant **Fomalhaut**) are now fully visible.

August

August – Introduction

Meteors

August is the month when one of the best meteor showers of the year occurs: the **Perseids**. This is a long shower, generally beginning about 17 July and continuing until around 24 August. The peak of the shower occurs over the night of 12 to 13 August, when the rate may reach as high as 100 meteors per hour (and on rare occasions, even higher). In 2026, there is a New Moon at the maximum, so there will be no competition from moonlight. The Perseids are debris from Comet 109P/Swift-Tuttle (the Great Comet of 1862). Perseid meteors are fast and many of the brighter ones leave persistent trains. Some bright fireballs also occur during the shower, arising from larger chunks of cometary debris.

In the southern hemisphere, the **Piscis Austrinid** shower continues until about 10 August, after maximum on 28 July. The **Southern δ-Aquariids** reach maximum of up to 25 meteors per hour on 30 July. The α-**Aurigid** shower begins on 28 August and is a short shower, lasting until 5 September and reaching maximum in 2026 on 1 September.

Comets

Comet 10P/Tempel 2 appears in June in Aquila at mag. 9, visible after 23:00. It moves into Capricornus in July and August, brightening to mag. 7 at point of closest approach on 3 August. Comet 10P/Tempel 2 has a period of 5.4 years. It is visible at dawn from central USA and Sydney; however, the comet is very low on the horizon from London.

Scientists working on gravity data for Mars and seismic waves measured by NASA's **InSight Lander** have revealed that there could be significant quantities of water trapped in rocks 11.5–20 km beneath the Martian surface. The surface of Mars once held lakes, rivers and oceans more than 3 billion years ago, and some of this water could be stored underground. NASA's mantra of 'follow the water' suggests conditions on Mars could have been conducive to simple life forms. The search for extraterrestrial life continues.

The planets

Mercury begins in *Gemini*, visible at dawn. A week later it enters *Cancer* and ends the month in *Leo*. After 27 August it trails the Sun, appearing shortly after sunset. Its magnitude brightens from 0.3 to –1.9 (at solar conjunction) and then dims to –1.6. It reaches western elongation on 2 August (mag. 0.2). On 15 August Mercury (mag. –1.2) is 0.5°N of *Jupiter* (mag. –1.8). *Venus* moves from Leo into *Virgo* on 1 August (mag. –4.3 to –4.6). On 15 August Venus is at eastern elongation (mag. –4.4). *Mars* starts in *Taurus* and swings into Gemini in the middle of the month; it brightens from 1.4 to 1.2 and dims slightly to 1.3 at the end of the month. Mars appears 1.5°N of *(1) Ceres* on 9 August in Taurus. Jupiter can be seen just before sunrise in Cancer (mag. –1.8). *Saturn* is in *Pisces*, visible after 23:00 (mag. 0.6 to 0.5). *Uranus* lies in Taurus, above the horizon from around midnight (mag. 5.8 to 5.7). *Neptune* is in Pisces and visible after 22:00 (mag. 7.7).

A

Sunrise and sunset

City	Date	Sunrise	Sunset
Buenos Aires, Argentina			
	Aug. 01	10:47	21:13
	Aug. 31	10:14	21:35
Cape Town, South Africa			
	Aug. 01	05:39	16:07
	Aug. 31	05:06	16:28
London, UK			
	Aug. 01	04:24	19:49
	Aug. 31	05:11	18:49
Los Angeles, USA			
	Aug. 01	13:04	02:54
	Aug. 31	13:26	02:20
Nairobi, Kenya			
	Aug. 01	03:37	15:41
	Aug. 31	03:31	15:36
Sydney, Australia			
	Aug. 01	20:48	07:15
	Aug. 31	20:15	07:36
Tokyo, Japan			
	Aug. 01	19:49	09:46
	Aug. 31	20:12	09:10
Washington, DC, USA			
	Aug. 01	10:09	00:19
	Aug. 31	10:36	23:40
Wellington, New Zealand			
	Aug. 01	19:29	05:26
	Aug. 31	18:48	05:55

NB: the times given are in Universal Time (UT)

The Moon's phases and ages

Northern Hemisphere

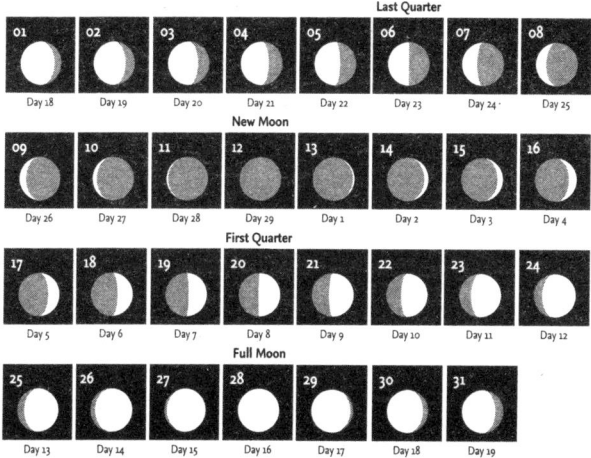

Southern Hemisphere

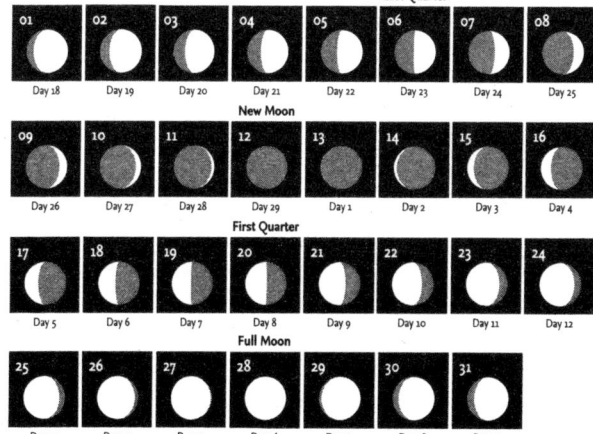

The Moon

The Moon in August
The Moon is Last Quarter on 6 August; the following day it passes 1.2°N of the **Pleiades**. On 9 August **Mars** (mag. 1.3) is 4.4°S of the Moon; a day later the waning crescent Moon is 3.6°S of **Pollux**. On 11 August **Mercury** (mag. −0.9) is 2.1°S of the thin crescent Moon. A partial solar eclipse will be visible from London on 12 August at 17:46; at maximum obscuration 91.4 per cent of the Sun will be covered by the New Moon. On 16 August **Venus** (mag. −4.4) lies 2.1°N of the waxing crescent Moon in the early evening. The following day the Moon approaches **Spica**, the star is 2.4°N. On 20 August the Moon is First Quarter; a day later it passes 0.6°S of **Antares**. On 28 August a partial lunar eclipse will be visible low on the western horizon from the UK, maximum coverage occurs at 04:13.

Solar eclipse of 12 August and partial lunar eclipse of 28 August
A total solar eclipse takes place on 12 August, visible from Iceland, Spain, Greenland and the Arctic. The maximum (totality) occurs at 17:46, lasting 2 minutes 18 seconds. A partial solar eclipse will be visible from Europe, west Africa and North America. A partial lunar eclipse on 28 August will be visible from Europe, Africa and the eastern Pacific. The time of maximum eclipse will be 04:13; the umbral (partial) eclipse will last 3 hours 18 minutes.

NASA mathematician: Katherine Johnson
Born on 26 August 1918 in West Virginia, Katherine Johnson was the third African American to gain a PhD in mathematics. She worked at NASA from 1953–1986, joining the all-black West Area Computing section at the National Advisory Committee for Aeronautics (NACA). Her first project required her to analyse data from flight tests and investigate the effects of turbulence on a plane crash.

Johnson moved into space technology in 1957, when NACA became NASA. She provided mathematical analysis of trajectories for the *Freedom 7* mission piloted by Alan Shepard in 1961, the first time an American entered space. In 1962, she checked orbital calculations carried out by early computers for an orbital mission called *Friendship 7* captained by astronaut John Glenn, the first American to orbit the Earth. Glenn had specifically requested Johnson to manually check the numbers, asking NASA engineers to 'Get the girl.' After Johnson confirmed the figures, he responded with 'If she says they're good, then I'm ready to go.'

Johnson contributed to the success of the Apollo missions; her trajectory calculations were used for the *Apollo 11* flight in 1969. She also worked on the *Apollo 13* mission in April 1970, planning an emergency safe return flight for the crew following the explosion of an oxygen tank. She later worked on the Space Shuttle programme and plans for a mission to Mars. She was awarded the Presidential Medal of Freedom by President Barack Obama in 2015, and in 2016 NASA named their Computational Research Facility in Virginia in her honour. She died four years later at the age of 101.

Katherine Johnson at NASA in 1966.

The Apollo 11 *mission on the launchpad, 16 July 1969.*

The Apollo 13 *crew returns safely to the South Pacific, 17 April 1970.*

Calendar for August

02	08:00	Mercury at greatest elongation (19.5°W, mag. 0.2)
06	02:21	Last Quarter Moon
07	06:23	Pleiades 1.2°S of the Moon
09	05:31	Mars (mag. 1.3) 4.4°S of the Moon
10	11:18	Moon at perigee = 363,288 km
10	22:38	Pollux 3.6°N of the Moon
11	12:48	Mercury (mag. −0.9) 2.1°S of the Moon
12	17:37	New Moon
12	17:46	Total solar eclipse (Iceland, Spain, Greenland and the Arctic). Partial eclipse visible from North America, west Africa and Europe.
12–13		Perseid meteor shower maximum
15	06:00	Venus at greatest elongation (45.9°E, mag. −4.4)
16	08:47	Venus (mag. −4.4) 2.1°N of the Moon
17	11:49	Spica 2.4°N of the Moon
20	02:46	First Quarter Moon
21	04:18	Antares 0.6°N of the Moon. An occultation will be visible from Antarctica, southern Argentina, Chile and Falkland Islands.
22	08:20	Moon at apogee = 404,644 km
28	04:13	Partial lunar eclipse. Visible from Europe (including London), Africa and the eastern Pacific.
28	04:18	Full Moon

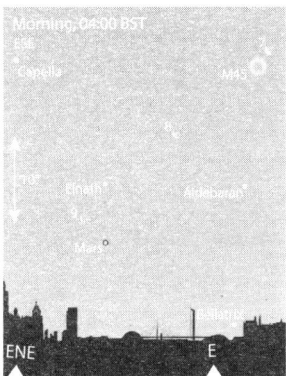

7–9 August: The waning crescent Moon from 7 August, moving past the Pleiades (M45) and Mars (as seen from London).

12 August: Partial solar eclipse as seen from London, UK. Maximum cover (91%) at 18:13 UTC.

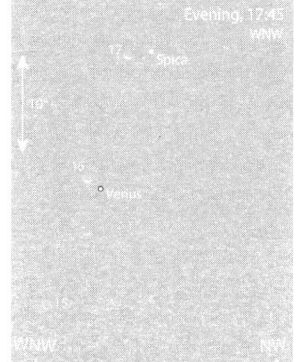

15–17 August: The waxing crescent Moon and Venus at greatest eastern elongation on 15 August. Moon next to Spica on 17 August (as seen from Sydney).

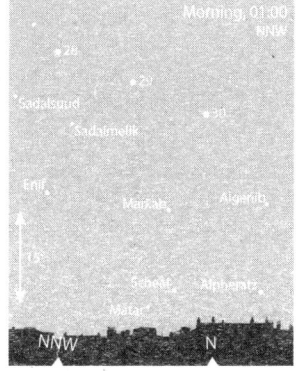

29 August: Full Moon in Aquarius, close to Sadelmelik (α Aquarii) and stars of Pegasus due north (as seen from Sydney).

A

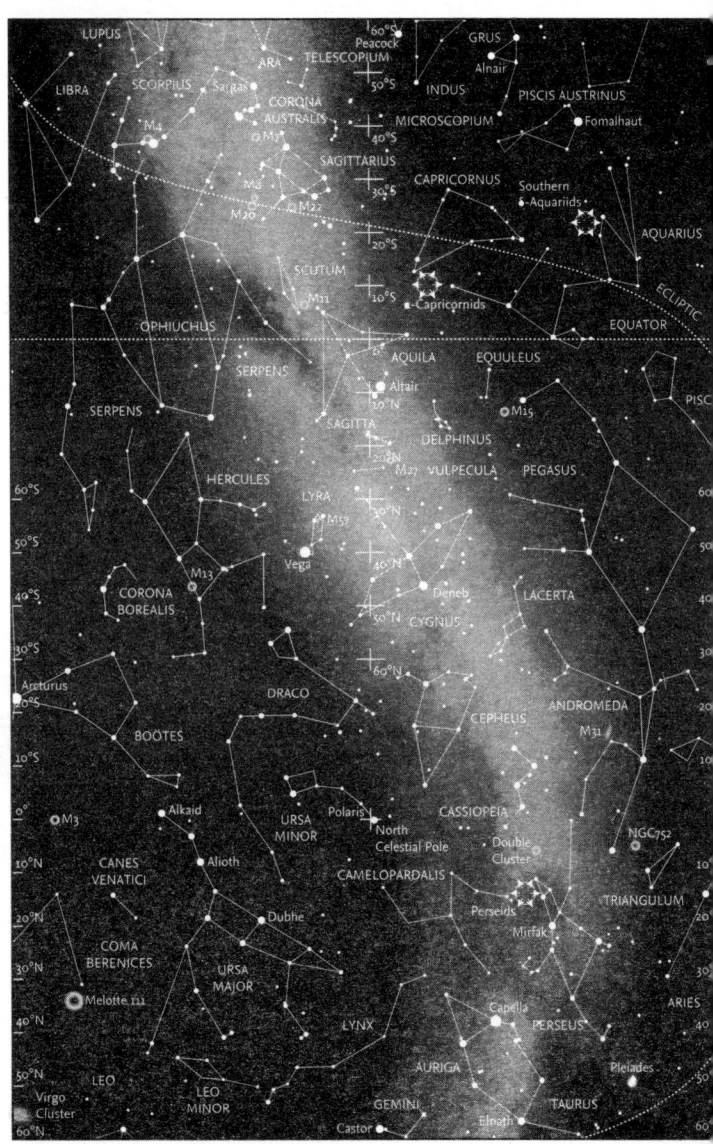

August – Looking North

Ursa Major is now the 'right way up' in the northwest, although some of the fainter stars in the south of the constellation are difficult to see. Beyond it, *Boötes* stands almost vertically in the west, but pale orange *Arcturus* is sinking towards the horizon. Higher in the sky, both *Corona Borealis* and *Hercules* are clearly visible, along with *Draco*.

For most mid-latitude northern observers, *Capella* is clearly seen in the northeast, but most of *Auriga* still remains below the horizon. Higher in the sky, *Perseus* is gradually coming into full view and, later in the night and later in the month, the beautiful *Pleiades* cluster rises above the northeastern horizon. Perseus straddles a narrow portion of the Milky Way and although not particularly distinct, a significant section of the Milky Way actually runs through Auriga in the direction of *Gemini*. Auriga contains numerous open clusters and a few emission nebulae, but is without globular clusters. Beyond Perseus in the east lies *Andromeda* and the small constellation of *Triangulum*. Between Perseus and *Polaris* lies the faint constellation of *Camelopardalis*.

The Milky Way is now running up the eastern side of the sky, where both *Cassiopeia* and *Cepheus* are well placed for observation, despite the fact that Cassiopeia is completely immersed in the band of the Milky Way, as is the 'base' of Cepheus. For observers south of the equator, these constellations are close to the horizon and *Ursa Minor* is on the horizon.

Pegasus and *Andromeda* are now well above the eastern horizon and, below them, the constellation of *Pisces* is climbing into view. For observers at about 40°N, both *Lyra* and *Cygnus*, with their brilliant principal stars, *Vega* and *Deneb*, respectively, are high overhead, near the zenith. The third star marking the other apex of the Summer Triangle, *Altair* in *Aquila*, is much farther to the south, and all three stars are best seen when looking south.

A

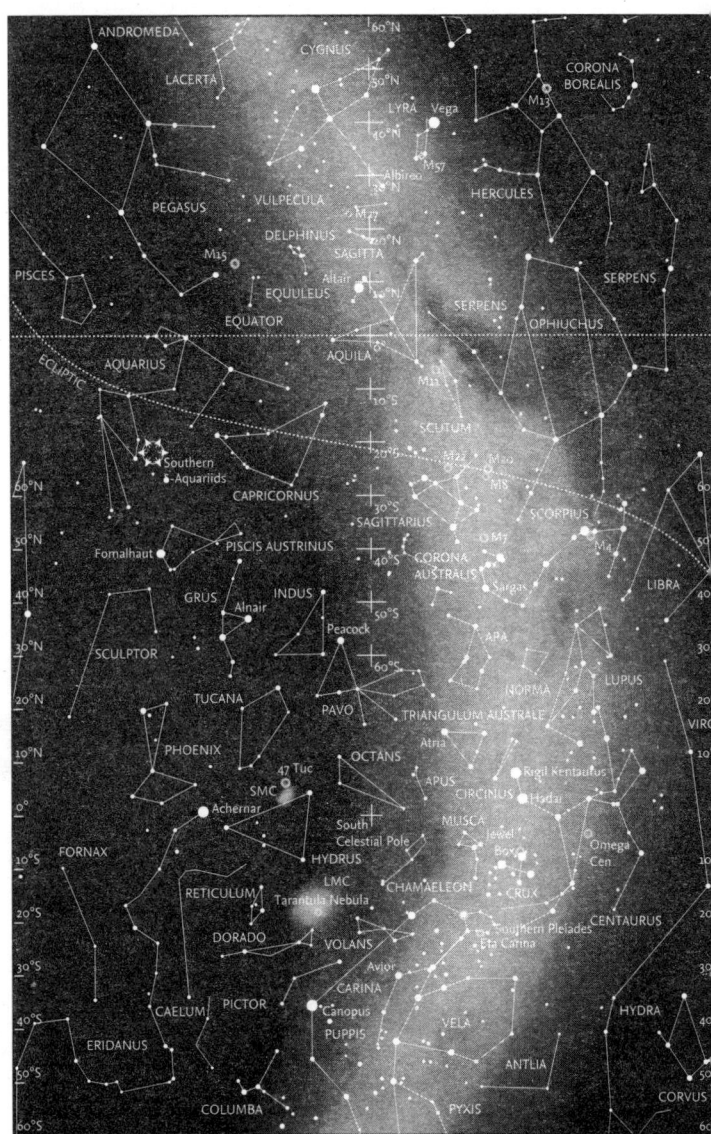

August – Looking South

The whole of the summer Milky Way stretches across the sky
in the south, from **Cygnus**, high in the sky near the zenith, past
Aquila, with bright **Altair** (α Aquilae), to part of the constellation of
Sagittarius close to the horizon, where the pattern of stars known
as the Teapot is visible. This area contains many nebulae and both
open and globular clusters. The star-forming Lagoon Nebula, **M8**, is
close to γ Sagittarii and it is visible to the naked eye on a clear dark
night. **M20**, the Trifid Nebula, is close to M8 and can be seen with
a small telescope. The **Great Dark Rift** marks the location of dense
clouds of dust that obscure starlight. To the southeast of Aquila lie
the two zodiacal constellations of **Capricornus** and **Aquarius.**

For observers at northern latitudes, the constellations of
Hercules and **Pegasus** are visible, together with **Delphinus** on
the east (one of the few constellations that actually bears some
resemblance to the creature after which it is named). Between
Albireo (β Cygni) and Altair lie the two small constellations of
Vulpecula and **Sagitta**. Below Aquila, mainly in the star clouds of the
Milky Way, lies **Scutum**, most famous for the bright open cluster
M11 or the Wild Duck Cluster. The two zodiacal constellations of
Scorpius and Sagittarius are still clearly visible, although becoming
rather low for observers at mid-northern latitudes, with even the
bright red supergiant star of **Antares** in Scorpius becoming difficult
to see. North of Scorpius is the large constellation of **Ophiuchus**,
with **Serpens Cauda** lying in front of the Great Dark Rift.

For observers farther south, three constellations are on
the meridian: **Volans**, **Chamaeleon** and **Octans**, with **Pavo** the
Peacock higher towards the zenith. Much of **Carina** is below the
horizon, although the **Eta Carinae Nebula** and the **Southern Pleiades**
are still visible. The **Large Magellanic Cloud** (LMC) and the faint
constellation of **Mensa** are slightly higher in the sky. Farther
south, **Lupus**, **Centaurus** and **Crux** are descending towards
the horizon. **Achernar**, the **Small Magellanic Cloud** (SMC) and
Hydrus are now well clear of the horizon, but below them are
more small constellations: **Dorado**, **Reticulum** and **Horologium.**
More of the river of **Eridanus** is visible, together with parts of
Pictor and **Fornax**. Much of **Cetus** has risen and the whole of the
western arm of **Pisces** is now clearly seen. **Sculptor**, **Grus** and
Piscis Austrinus lie halfway between the eastern horizon and the
zenith, with faint **Microscopium** closer to the zenith.

A

September

September – Introduction

The (northern) autumnal equinox occurs on 23 September, when the Sun moves south of the equator in Virgo.

Meteors
There are two minor showers in September visible in the northern hemisphere. The *α-Aurigids* starts on 28 August and ends on 5 September and tends to have two peaks of activity. The principal peak occurs on 1 September. In 2026, the Moon will be a bright waning gibbous, so conditions are unfavourable during the shower. At maximum, the hourly rate hardly reaches 10 meteors per hour, although the meteors are bright and relatively easy to photograph. The *Southern Taurid* shower begins on 10 September and, although rates are low, often produces very bright fireballs. This is a very long shower, lasting until about 20 November.

There is one minor shower in the southern hemisphere, the *Piscids*, active throughout September, with an activity of 3–5 meteors per hour. As a slight compensation for the lack of shower activity, in September the number of sporadic meteors reaches its highest rate of the year, although these are completely unpredictable in location, direction and magnitude.

The planets
Mercury starts in *Leo* and enters *Virgo* by the end of the first week of September, visible after sunset (mag. −1.5 to −0.1). On 26 September it moves 0.8°N of *Spica*, at mag. −0.2. *Venus* sits in Virgo for the month, approaching the Sun and visible in the early evening (mag. −4.6 to −4.8). On 1 September it is 1.2°S of Spica. *Mars* starts the month in *Gemini* and slowly moves into *Cancer*. The red planet is visible from around 02:00 until sunrise (mag. 1.2 to 1.1). *Jupiter* can be observed a few hours before sunrise in Cancer, crossing into Leo near the end of the month (mag. −1.8 to −1.9). *Saturn* lies on the border between *Pisces* and *Cetus*, visible after 21:00 (mag. 0.5 to 0.3). *Uranus* stays in *Taurus*, visible in the late evening (mag. 5.7). It enters retrograde motion on 10 September. *Neptune* is in Pisces (mag. 7.7), reaching opposition on 26 September, a distance of 28.9 AU from Earth.

A large, hot, Jupiter-sized exoplanet 637 light-years away has iron winds on its day side and iron rain on its night side. The planet, called **WASP-76 b**, has a temperature of over 2,000°C and is tidally locked with its host star – one side of the planet continuously faces the star. Mapping of the winds was achieved with an instrument called a spectrograph on the European Southern Observatory's Very Large Telescope in Chile.

Space dust and the Earth's climate

Astronomers from Boston University in the USA have used a 2D atmospheric chemistry model to show that our Solar System passed through dense interstellar clouds between 2 and 7 million years ago, with an ensuing ice age. The interstellar medium is the gas (mostly hydrogen and helium) and dust distributed between stars in galaxies. Interstellar clouds vary in temperature and density; the coldest and densest clouds are the sites of star formation. The chemistry of the clouds evolves over millions and billions of years as stars die and the elements formed in their cores are recycled for the next stellar generation.

The Solar System is in a 230 million-year orbit of the centre of the Milky Way, and occasionally we pass through dense interstellar clouds along the way. At two points in our past these clouds compressed the solar wind (the heliosphere), thus exposing Earth's atmosphere to the interstellar medium. Our journeys through these clouds lasted around 100,000 years and they had a significant impact on our climate; an ice age was triggered due to the formation of global noctilucent clouds (containing water ice crystals) in the upper atmosphere, reducing the amount of sunlight reaching the surface of the Earth by 7 per cent.

The next step in the team's research is to create a 3D computational model to help uncover the global response to these cloud crossings.

S

Sunrise and sunset

City	Date	Sunrise	Sunset
Buenos Aires, Argentina			
	Sep. 01	10:13	21:35
	Sep. 30	09:32	21:56
Cape Town, South Africa			
	Sep. 01	05:05	16:29
	Sep. 30	04:25	16:49
London, UK			
	Sep. 01	05:13	18:47
	Sep. 30	05:59	17:41
Los Angeles, USA			
	Sep. 01	13:27	02:18
	Sep. 30	13:47	01:38
Nairobi, Kenya			
	Sep. 01	03:30	15:35
	Sep. 30	03:19	15:26
Sydney, Australia			
	Sep. 01	20:14	07:37
	Sep. 30	19:34	07:57
Tokyo, Japan			
	Sep. 01	20:13	09:09
	Sep. 30	20:35	08:27
Washington, DC, USA			
	Sep. 01	10:37	23:38
	Sep. 30	11:03	22:52
Wellington, New Zealand			
	Sep. 01	18:47	05:56
	Sep. 30	17:57	06:25

NB: the times given are in Universal Time (UT)

The Moon's phases and ages

Northern Hemisphere

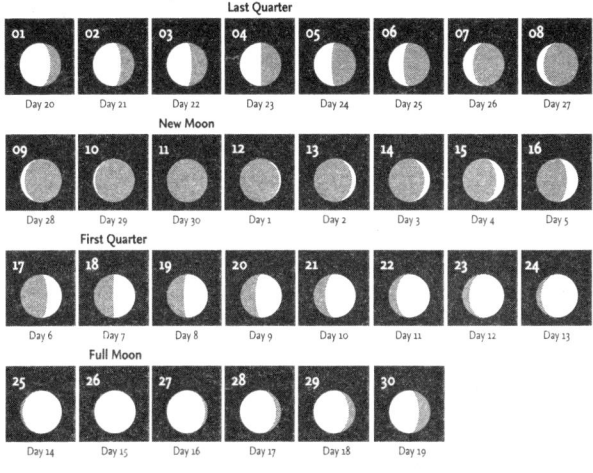

Southern Hemisphere

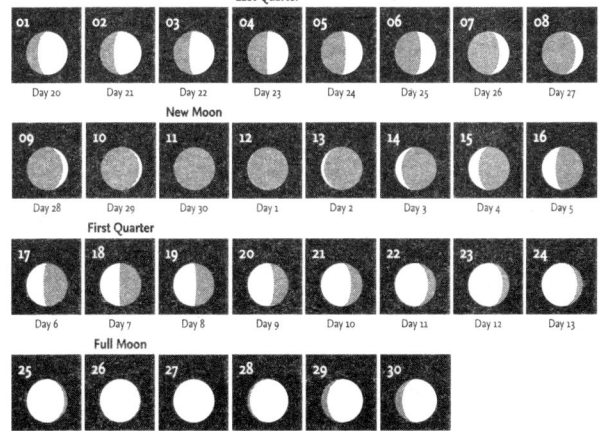

The Moon

The Moon in September

On 3 September the waning gibbous Moon sits 1.2°N of the *Pleiades*. The following day it reaches Last Quarter. On 6 September *Mars* (mag. 1.2) is 3.0°S of the waning crescent Moon. A day later it moves 3.6°S of *Pollux*. On 8 September the Moon moves 0.8°N of *Jupiter* (mag. −1.8); the following day the thin crescent Moon is 0.5°S of *Regulus*. The New Moon takes place on 11 September; two days later the faint waxing crescent Moon is 2.4°S of *Spica*. *Venus* (mag. −4.8) is 0.5°S of the Moon on 14 September, when an occultation will occur in the morning, lost in the daylight. On 17 September *Antares* lies 0.6°N of the waxing crescent Moon, followed by a First Quarter Moon a day later. A Full Moon occurs on 26 September and the Moon rejoins the Pleiades on 30 September, lying 1.1°N of the star cluster.

Planets orbiting dead stars

Research published in September 2024 in *Nature* shows a planetary system orbiting a white dwarf star – the same type of star our Sun will eventually become in around 6 billion years' time. The system lies 4,000 light-years from Earth. The white dwarf has around half the mass of the Sun, and a planet around the size of the Earth, in an orbit twice as wide as the Earth's orbit of the Sun. There is also an object with 17 times the mass of Jupiter, thought to be a brown dwarf (a failed star).

The system was detected using a method known as gravitational microlensing in 2020. The planetary system lies between the Earth and a background star (24,000 light-years away). Its gravity acts as a lens brightening the far-away starlight as it passes though the planetary system on its way to our telescopes. The planetary system provides insight into the future for our Solar System when the Sun evolves into a red giant, sheds its outer layers forming a planetary nebula, and ends its life as a white dwarf, slowly radiating the last of its energy into the cold vacuum of space.

Meteorites: Messengers of the sky

Among the planets, asteroids and comets are an unimaginable number of smaller space rocks and debris called meteoroids. Around 25 million meteoroids and smaller objects enter Earth's atmosphere each day. They can travel through the atmosphere at speeds reaching 70 km per second; they can flare up as they react with the atmosphere and are known as meteors if they are visible in the night sky. If they survive their journey to the ground they become known as meteorites. These are easier to find in desert areas and icy regions such as Antarctica.

Hasnaa Chennaoui Aoudjehane is a Moroccan planetary scientist at Hassan II University, Casablanca, Morocco. She has an asteroid named after her, **asteroid 299020 Chennaoui,** in recognition of her efforts to promote geology and meteoritics in North Africa and the Middle East. In 2014, she organized the 77th Meteoritical Society Meeting in Casablanca, the first such meeting in an Arabic country. Chennaoui Aoudjehane led the project characterizing the Martian meteorite called Tissint, which fell in Morocco 48 km away from a town called Tissint on 18 July 2011. The meteorite left the Martian surface close to a million years ago; it is thought to have originated from deep underground and shows weathering by water.

S

Calendar for September

01		α-Aurigid meteor shower maximum
01	13:24	Venus (mag. –4.6) 1.2°S of Spica
03	12:03	Pleiades 1.2°S of the Moon
04	07:51	Last Quarter Moon
06	18:24	Mars (mag. 1.2) 3.0°S of the Moon
06	20:26	Moon at perigee = 368,255 km
07	06:32	Pollux 3.6°N of the Moon
08	18:13	Jupiter (mag. –1.8) 0.8°S of the Moon.
09	19:36	Regulus 0.5°N of the Moon
11	03:27	New Moon
13	20:53	Spica 2.4°N of the Moon
14	11:10	Venus (mag. –4.8) 0.5°S of the Moon. Occultation visible from Asia, Africa, Europe (including London) and western Russia.
17	12:18	Antares 0.6°N of the Moon. Occultation visible from Antarctica, Australia and Tasmania.
18	20:44	First Quarter Moon
19	03:00	Moon at apogee = 404,217 km
23	00:06	Autumnal Equinox
26	01:49	Mercury (mag. –0.2) 0.8°N of Spica
26	16:49	Full Moon
30	17:39	Pleiades 1.1°S of the Moon

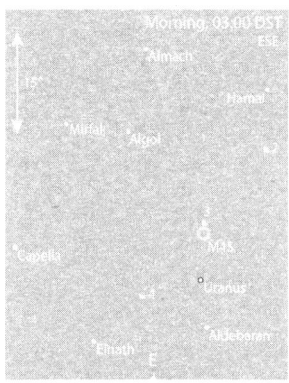

3 September: *Waning gibbous Moon next to the Pleiades (M45), Uranus nearby, along with Aldebaran, Capella and Elnath (as seen from central USA).*

7 September: *Waning crescent Moon approaching Mars in Gemini, Pollux and Castor close by (as seen from Sydney).*

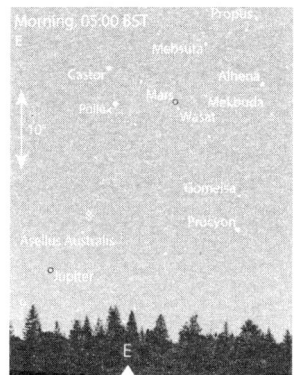

9 September: *Waning crescent Moon near Jupiter and Asellus Australis (δ Cnc) in Cancer lying due east (as seen from London).*

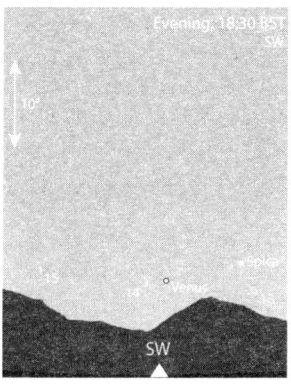

14 September: *Waxing crescent Moon and Venus just after sunset. An occultation takes place at 11:10 UTC (as seen from London).*

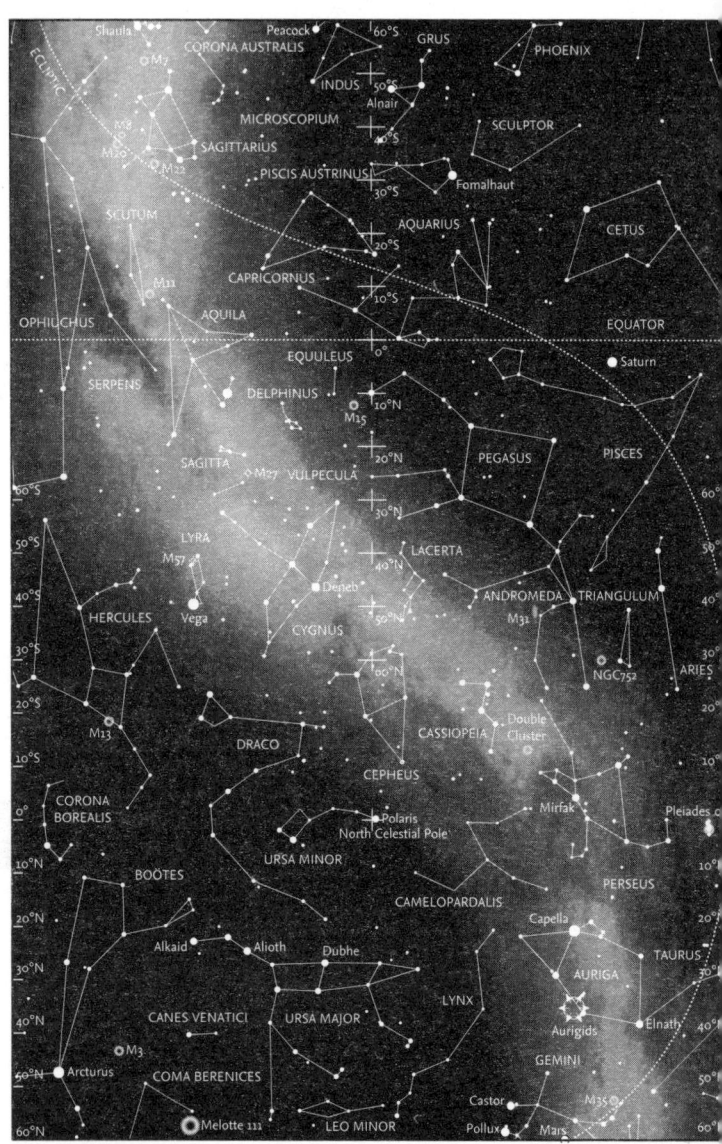

September – Looking North

For mid-northern observers, **Ursa Major** is now 'right way up' low in the north and to the northwest **Arcturus** and much of **Boötes** sink below the horizon later in the night and later in the month. In the northeast, **Auriga** and its dazzling star **Capella** are beginning to climb higher in the sky. Later in the month, **Taurus**, with orange **Aldebaran** (α Tauri), and even **Gemini**, with **Castor** and **Pollux**, become visible in the east and northeast later in the night. Due east, **Andromeda** is now clearly visible, with the small constellations of **Triangulum** and **Aries** (the latter a zodiacal constellation) directly below it. Practically the whole of the northern Milky Way is visible, arching across the sky, both in the north and in the south. The clouds of stars are easier to see in **Cassiopeia** and **Cygnus**. The **Double Cluster** in **Perseus** is well placed for observation.

The clouds of stars forming the Milky Way are not particularly striking in Auriga and Perseus, but beyond Cassiopeia, and on towards Cygnus, they become much denser and easier to see. Cygnus is high in the northwest and **Deneb**, its principal star, is close to the zenith for observers at 40–50° north, with **Lyra** and bright **Vega** slightly farther west. Farther towards the south, most of **Hercules** is clearly visible, with the Keystone and the globular cluster **M13**. **Cepheus** is 'upside down' near the zenith, apparently hanging from the Milky Way. The head of **Draco** lies between Hercules and Cepheus and the whole of that constellation is easily seen as it curls around **Ursa Minor** and the North Celestial Pole.

S

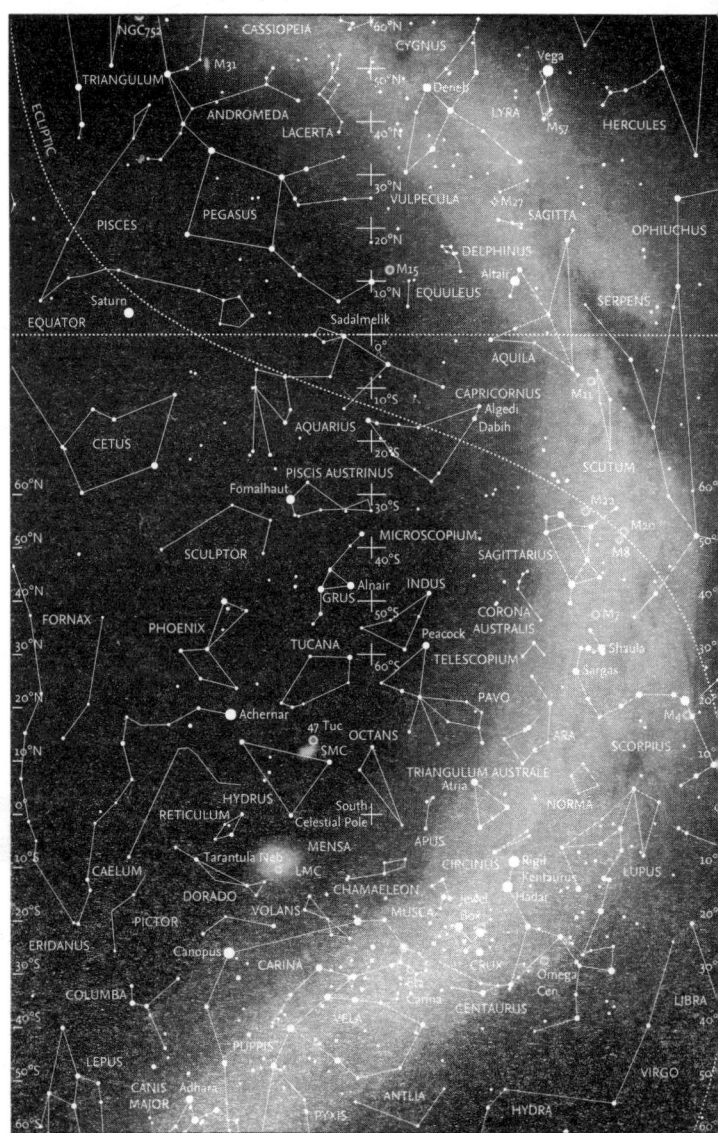

September – Looking South

The Summer Triangle is now high in the southwest, with the Great Square of **Pegasus** high in the southeast. The **Great Dark Rift** is clearly visible, starting near **Deneb** and running down the centre of the Milky Way towards the centre of the Galaxy in **Sagittarius**, and beyond into **Scorpius**, only petering out in **Centaurus** and **Crux**. Below Pegasus are the two zodiacal constellations of **Capricornus** and **Aquarius**. **Algedi** (α Capricorni) is actually an optical double, with the two stars (α¹ Cap and α² Cap) readily seen with the naked eye. **Dabih** (β Capricorni), just to the south, is also a double star and the components are relatively easy to separate with binoculars. In Aquarius, just to the east of **Sadalmelik** (α Aquarii) there is a small asterism consisting of four stars, resembling a tiny letter 'Y', known as the Water Jar. Below Aquarius is a sparsely populated area of the sky with just one bright star, **Fomalhaut** (alpha Piscis Austrini), in the constellation of **Piscis Austrinus**. Below it are the constellations of **Grus** and **Pavo**, with **Achernar** (α Eridani) and almost the whole of the long river constellation of **Eridanus**.

Another zodiacal constellation, **Pisces**, is now clearly visible to the east of Aquarius. Although faint, there is a distinctive asterism of stars, known as the Circlet, south of the Great Square. Still farther down towards the horizon is the constellation of **Cetus**, with the famous variable red giant star **Mira** (o Ceti) at its centre. When Mira is at maximum brightness (around mag. 3.5) it is clearly visible to the naked eye.

For observers south of the equator, the smallest of the constellations, **Crux**, is now very low, but **Canopus** (α Carinae) has now become visible once more and is hugging the horizon for observers at the latitude of Sydney in Australia. **Dorado** and the **Large Magellanic Cloud** (LMC) are higher and easier to see, as are the faint constellations of **Pictor**, **Caelum** and **Reticulum**. The **Small Magellanic Cloud** (SMC) and **47 Tucanae** (the globular cluster in **Tucana**) are now nearly halfway between the horizon and the zenith. Although becoming low, the whole of **Centaurus** and **Lupus** remain visible in the southwest. **Scorpius**, with red-orange **Antares**, is beginning to descend in the west, but **Sagittarius** is clearly seen high in the sky. Between the 'sting' of Scorpius and those two bright stars are the small constellations of **Ara** and **Triangulum Australe**.

October

October – Introduction

Towards the end of the month (Sunday 25 October) Summer Time ends in Europe, with the UK reverting to Greenwich Mean Time and Europe to Central European Time. Daylight Saving Time starts where used in Australia and in New Zealand on Sunday 4 October.

Meteors

The **Orionids** meteor shower starts on 2 October and builds up to a broad maximum lasting around a week centred on 21 October. Like the May *η-Aquariid* shower, the Orionids are associated with Comet 1P/Halley. During this second pass through the stream of particles from the comet, slightly fewer meteors are seen than in May, but conditions are more favourable for northern observers. In both showers the meteors are very fast and many leave persistent trains, with hourly rates around 15. In 2026, there is a waxing gibbous Moon lighting up the sky and interfering with observations.

The faint shower of the **Southern Taurids** (often with bright fireballs) peaks on 10 October when the Moon is New and observing conditions are favourable. Towards the end of the month (around 20 October), another shower (the **Northern Taurids**) begins to show activity, peaking in November. The parent comet for both Taurid showers is Comet 2P/Encke. The meteors in both Taurid streams are relatively slow and bright. A minor shower of short duration, the **Draconids**, is visible to northern observers. The shower begins on 6 October and peaks on 9 October when skies are dark as the thin waning crescent Moon rises just before dawn.

The planets

Mercury begins the month in **Virgo** and swiftly moves into **Libra**. It is visible after sunset and approaches the Sun after eastern elongation on 12 October (mag. −0.1 to 2.6). **Venus** is in Virgo visible in the early evening. It approaches the Sun and is seen at dawn after 24 October (mag. −4.8 to −4.1). **Mars** is in **Cancer**, it progresses into **Leo** for Halloween, approaching **Jupiter** along the ecliptic and visible after midnight (mag. 1.1 to 0.8). On 11 October Mars (mag. 1.1) sits next to the **Beehive Cluster, M44** (mag. 3.1), 1.0°N. Jupiter is in Leo, climbing above the horizon from 01:00 (mag. −1.9 to −2.0). **Saturn** lies on the border between **Pisces** and **Cetus**, visible after sunset (mag. 0.3 to 0.5). On 4 October it is at opposition (mag. 0.3), positioned 8.4 AU from Earth. **Uranus** is in

Taurus and visible in the late evening (mag. 5.7 to 5.6). *Neptune* is in Pisces and visible just after sunset (mag. 7.7).

Observing faint objects in the night sky

Our eyes use different types of cells during the day and at night. In daylight the cone cells are active and provide sensitivity to colour (wavelength of light), and in darkness rod cells switch on, allowing us to achieve low-level monochrome night vision. With 120 million rod cells and 7 million cone cells in the retina (a light-sensitive layer in the back of our eyes) rod cells offer heightened light sensitivity but lower resolution due to multiple inputs to a single nerve cell. This setup sacrifices detail for greater sensitivity to light.

Rod cells take time to become fully activated. During the day these sensitive photoreceptors are saturated with sunlight and are inactive. After sunset it can take up to 30 minutes for rod cells to fully regenerate the light-sensitive protein rhodopsin. This night vision can be destroyed by looking at a mobile phone, using an ordinary torch, streetlights or the headlamps of a car. It will also be lost by looking at a bright Moon in the sky. To maintain night vision when observing the night sky the use of a red torch is recommended. Mobile phone stargazing apps have a red night vision mode that should be switched on.

When observing dim celestial objects, including faint stars or galaxies like *Andromeda*, a technique known as averted vision is useful. Rod cells are located in the outer regions of the retina and therefore provide us with peripheral vision. By looking slightly to the side of an object at night, the rod cells are activated and the object appears brighter. Averted vision requires practice as there is a tendency to instinctively look directly at the object.

China launched its new *Long March-6* rocket in October 2024 carrying a batch of 18 satellites to join its **Thousand Sails** constellation. A total of 1,296 satellites will be placed in orbit by 2027, an initiative to rival that of Starlink owned by SpaceX. The satellites are in polar orbit, passing through the North and South Celestial Poles. An increasing number of satellites reflecting sunlight adds to the growing problem of light pollution for astronomers; currently there are no formal international regulations to protect our visibility of the night sky.

Sunrise and sunset

City	Date	Sunrise	Sunset
Buenos Aires, Argentina			
	Oct. 01	09:30	21:57
	Oct. 31	08:53	22:22
Cape Town, South Africa			
	Oct. 01	04:24	16:49
	Oct. 31	03:47	17:14
London, UK			
	Oct. 01	06:01	17:38
	Oct. 31	06:52	16:35
Los Angeles, USA			
	Oct. 01	13:48	01:37
	Oct. 31	14:12	01:01
Nairobi, Kenya			
	Oct. 01	03:19	15:26
	Oct. 31	03:12	15:21
Sydney, Australia			
	Oct. 01	19:33	07:58
	Oct. 31	18:56	08:22
Tokyo, Japan			
	Oct. 01	20:35	08:25
	Oct. 31	21:02	07:47
Washington, DC, USA			
	Oct. 01	11:04	22:51
	Oct. 31	11:34	22:09
Wellington, New Zealand			
	Oct. 01	17:56	06:26
	Oct. 31	17:09	07:00

NB: the times given are in Universal Time (UT)

The Moon's phases and ages

Northern Hemisphere

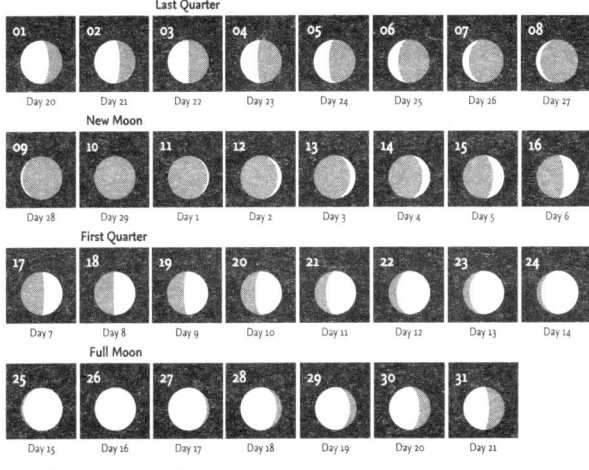

Southern Hemisphere

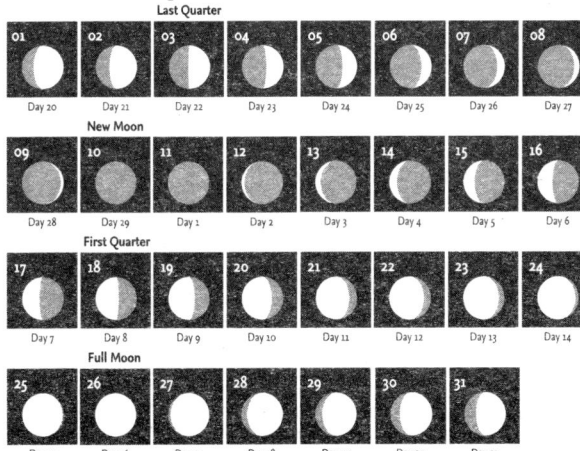

195

The Moon

The Moon in October

On 3 October the Last Quarter Moon dominates the late night sky. A day later the waning crescent Moon sits 3.8°S of *Pollux*. On 5 October *Mars* (mag. 1.1) sits 1.2°S of the Moon; a day later the Moon moves 0.2°N of *Jupiter* (mag. –1.9). On 7 October *Regulus* is 0.6°N of the faint crescent Moon and a New Moon takes place on 10 October. On 12 October the thin waxing crescent Moon is 3.1°N of *Venus* (mag. –4.5) and later that day it is 2.1°S of *Mercury* (mag. 0.0). Two days later the Moon joins *Antares*, sitting 0.4°S of the red star. On 18 October the Moon is First Quarter; a Full Moon occurs on 26 October. Two days later it lies 1.0°N of the *Pleiades* in *Taurus*. The waning gibbous Moon rejoins *Pollux* on 31 October, sitting 4.0°S of one of the heads of *Gemini*.

Organic molecule found in interstellar cloud

A carbon-based molecule called pyrene has been discovered in an interstellar cloud called the Taurus molecular cloud, *TMC-1*, located 430 light-years away. Pyrene belongs to a family of molecules called polycyclic aromatic hydrocarbons, or PAHs, and may be the source of much of the carbon in the Solar System. Pyrene has also been found in samples from the asteroid *Ryugu* taken by the Japanese spacecraft *Hayabusa2*, which landed on the asteroid in June 2018. PAHs can be detected by the infrared radiation they emit; however, it is difficult to accurately identify the molecules themselves. Radio astronomy using the Green Bank Observatory in West Virginia, USA, allowed the astronomers at the Massachusetts Institute of Technology to discover a version of pyrene called cyanopyrene. Their discovery supports the hypothesis that carbon locked in pyrene molecules is used throughout the evolution of a planetary system, from the nebulous birthplace to the rocks themselves, including asteroids, moons and planets.

Ejnar Hertzsprung: The life cycle of stars

At the turn of the twentieth century, Danish astronomer Ejnar Hertzsprung (1873–1967) was analysing observations of stars and categorizing them according to their physical and chemical properties. He was surprised to find red stars of similar surface temperature and spectral type with significantly different absolute magnitudes (true brightnesses). The brighter stars were red giants or supergiants – evolved stars reaching the end of their lives. In around 4–5 billion years, the Sun will expand into a red giant, with its surface reaching across the orbit of Earth and swallowing up Mercury and Venus. The change in colour is the result of a drop in surface temperature. A famous red supergiant star called Betelgeuse is visible to the naked eye in the constellation Orion. Sun-like stars gradually evolve into white dwarf stars; more massive stars like Betelgeuse will undergo a powerful supernova explosion, bright enough to outshine the entire galaxy in which the star resides. They end their lives either as incredibly dense neutron stars or black holes.

In 1911, Hertzsprung created a diagram to track the evolution of stars, further developed by Henry Norris Russell in 1913. The Hertzsprung–Russell diagram displays the absolute magnitude (intrinsic brightness) of stars against their spectral class or temperature. The life cycle of stars of different masses can be determined from the observations of numerous stars. This is analogous to uncovering the human life cycle by observing a large population of people of all ages, without the need for watching a human typically age from birth to death.

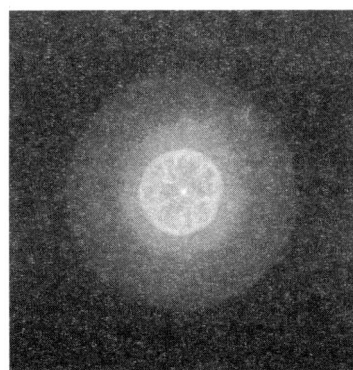

Planetary nebula IC 3568, the Lemon Slice Nebula, is an evolved star that has passed through its red giant phase. The star lies 4,500 light-years away in Camelopardalis.

O

Calendar for October

01	20:41	Moon at perigee = 369,338 km
03	13:25	Last Quarter Moon
04	12:21	Saturn (mag. 0.3) at opposition
04	12:27	Pollux 3.8°N of the Moon
05	05:30	Mars (mag. 1.1) 1.2°S of the Moon
06	10:18	Jupiter (mag. −1.9) 0.2°S of the Moon
07	02:57	Regulus 0.6°N of the Moon. Occultation visible from Africa.
10	15:50	New Moon
10		Southern Taurid meteor shower maximum
12	02:30	Venus (mag. −4.5) 3.1°S of the Moon
12	10:00	Mercury at greatest elongation (25.2°E, mag. 0.0)
12	20:08	Mercury (mag. 0.0) 2.1°N of the Moon
14	20:25	Antares 0.4°N of the Moon. Occultation visible from eastern Uruguay, southern Brazil, South Georgia and the South Sandwich Islands and Saint Helena.
16	22:56	Moon at apogee = 404,639 km
18	16:13	First Quarter Moon
21–22		Orionid meteor shower maximum
26	04:12	Full Moon
28	01:11	Pleiades 1.0°S of the Moon
28	18:01	Moon at perigee = 364,411 km
31	18:00	Pollux 4.0°N of the Moon

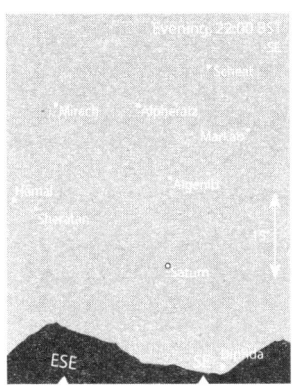

4 October: *Saturn in Pisces. Diphda, Alpheratz, Algenib and Markab nearby (as seen from London).*

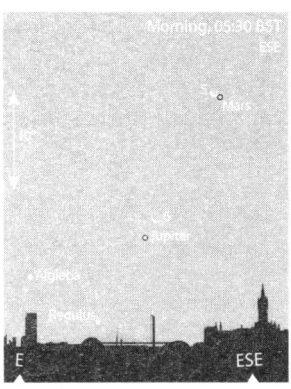

5 October: *Waning crescent Moon next to Mars, Jupiter sits nearby (as seen from London).*

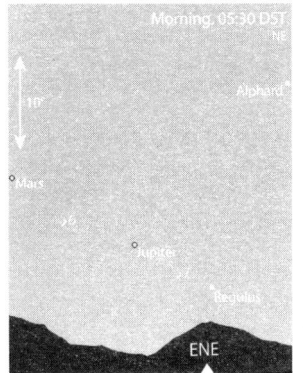

7 October: *Waning crescent Moon close to Jupiter and Regulus, Mars sits northwards (as seen from Sydney).*

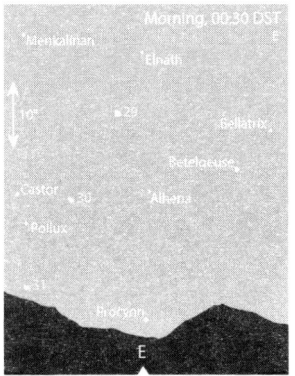

29–31 October : *Waning gibbous Moon close to Pollux and Castor; Betelgeuse and Procyon are nearby (as seen from central USA).*

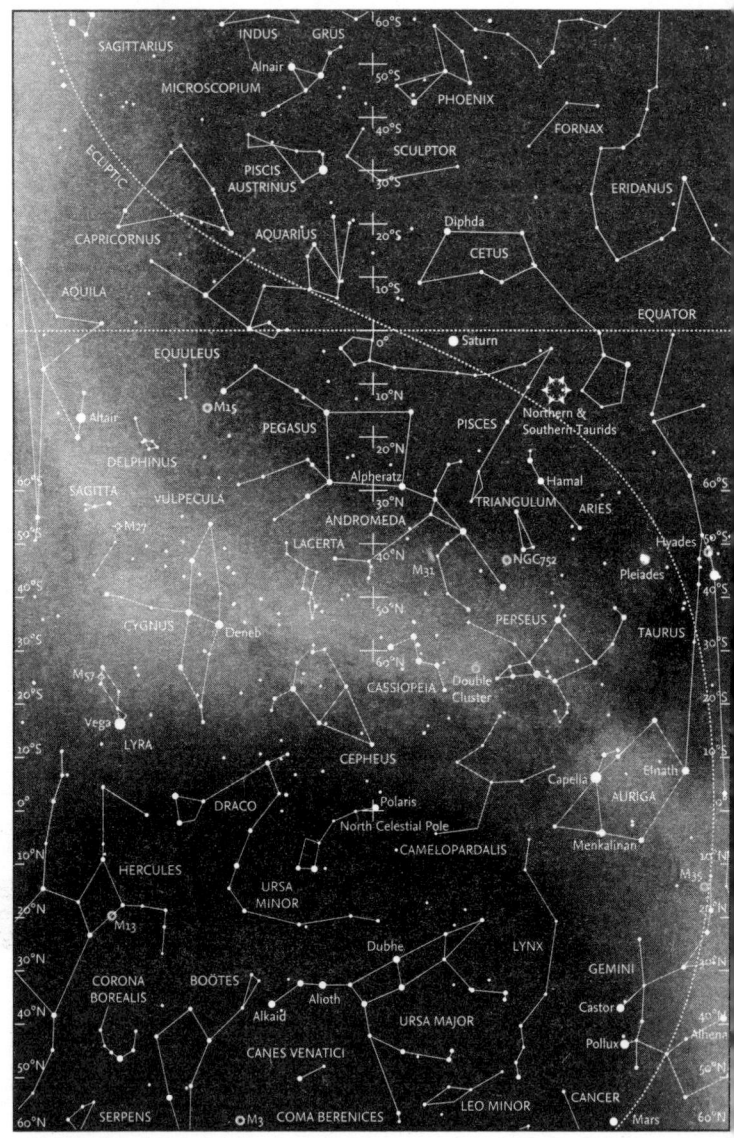

October – Looking North

The Milky Way arches across the northern sky, running from **Auriga** in the east to **Cygnus** and **Aquila** in the west. **Ursa Major** is grazing the horizon in the north, while high overhead are the constellations of **Cepheus**, **Cassiopeia** and **Perseus**, with the Milky Way between Cepheus and Cassiopeia near the zenith. In the east lies **Taurus** with the **Pleiades**, **Hyades** and orange **Aldebaran**. Later in the night, and later in the month, **Orion** and **Gemini** are starting to rise clear of the horizon.

The head of **Draco** is at about the same altitude as **Polaris** in **Ursa Minor**, as is **Capella** (α Aurigae). The large constellation of Ursa Major is now directly beneath the Pole, visible above the horizon to the north. **Castor** and **Pollux** (α and β Geminorum, respectively) are in the northeast, while the northernmost stars of **Boötes** and the circlet of **Corona Borealis** are in the northwest. The brightest star in Boötes, orange-tinted **Arcturus**, is below the horizon early in the night. **Hercules** is also descending towards the western horizon. In the northwest, the constellation of **Lyra** lies farther south, clear of the Milky Way. The three stars of the Summer Triangle are still clearly visible, although Aquila and **Altair** are beginning to approach the horizon in the west. The long chain of faint stars that is the constellation of **Lynx** runs almost vertically between Ursa Major and Gemini, with the other faint constellation of **Camelopardalis** between it and Perseus.

O

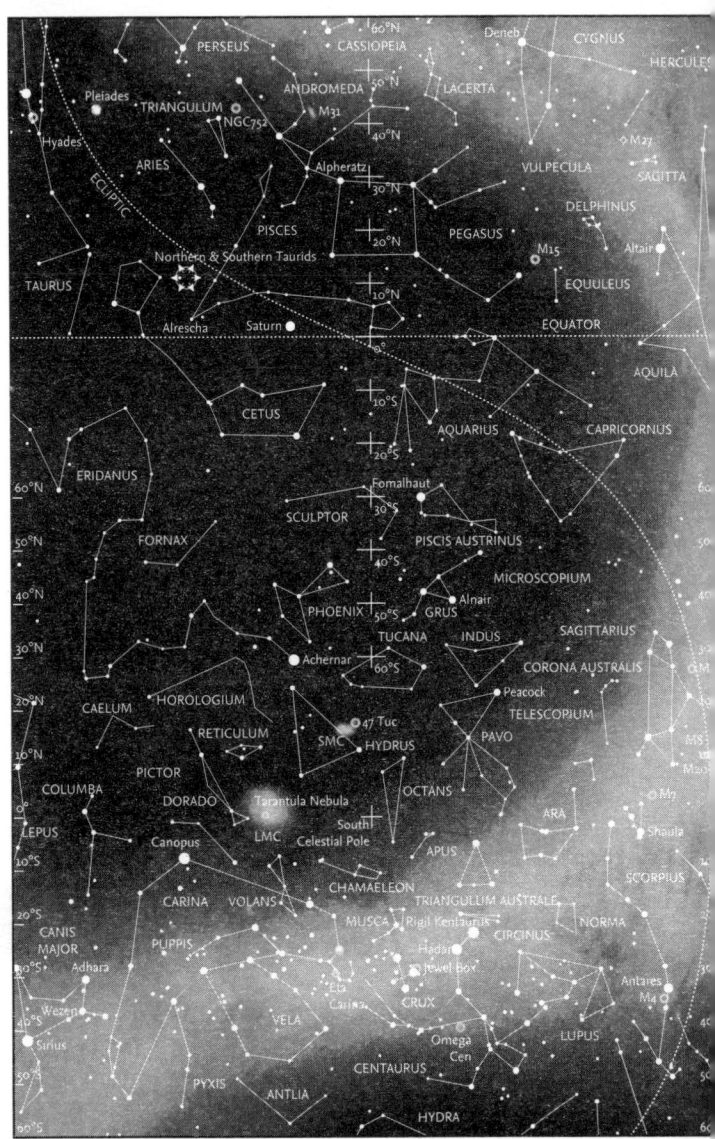

October – Looking South

The Great Square of **Pegasus** dominates the southern sky, framed by the two chains of stars that form the constellation of **Pisces**, together with **Alrescha** (α Piscium) at the point where the two lines of stars join. Also clearly visible is the constellation of **Cetus**, below Pegasus and Pisces. The star at the northeastern corner of the Great Square is actually **Alpheratz** (α Andromedae). The two parts of the zodiacal constellation of Pisces are to the south and east of Pegasus. Although **Capricornus** is beginning to disappear, **Aquarius** to its east is well placed in the south, with solitary **Fomalhaut** and the constellation of **Piscis Austrinus** (the Southern Fish) beneath it, close to the horizon.

The main band of the Milky Way and the **Great Rift** runs down from **Cygnus**, through **Vulpecula**, **Sagitta** and **Aquila** towards the western horizon. **Delphinus** and the tiny, unremarkable constellation of **Equuleus** lie between the band of the Milky Way and Pegasus. **Andromeda** is clearly visible high in the sky to the southeast, with the small constellation of **Triangulum** and the zodiacal constellation of **Aries** below it.

For observers at about 30° south (roughly the latitude of Sydney in Australia), the two brightest stars of **Centaurus** (**Rigil Kentaurus** and **Hadar**) are brushing the horizon, while the small constellation of **Crux** is lost below it. Only observers even farther south will be able to see that constellation in full, together with the constellation of **Lupus**, the full extent of Centaurus and the large constellation of **Vela**, with **Puppis** in the southeast. The **False Cross** has reappeared, also very low, farther towards the east on the opposite side of the meridian. Brilliant **Canopus** (α Carinae) is now much higher as is the **Large Magellanic Cloud** (LMC). The **Small Magellanic Cloud** (SMC) and the globular cluster **47 Tucanae** are on the southern meridian, halfway to the zenith, as is **Achernar** (α Eridani). The whole of the long constellation of **Eridanus** is now visible together with **Rigel** in **Orion.** Straddling the meridian slightly east of Piscis Austrinus is the faint constellation of **Sculptor.** Below (south of) these two constellations is the roughly cross-shaped constellation of **Grus** with, on the other side of the meridian, **Phoenix.** Lower in the sky lies the triangular constellation of **Hydrus**, and at about the same altitude towards the west is

O

the constellation of **Pavo** with **Peacock** (α Pavonis). **Ophiuchus** has now slipped below the horizon as has much of **Scorpius**, only the 'tail' of which remains visible. **Sagittarius** is getting lower, but remains visible, as do the zodiacal constellations of Capricornus and Aquarius. Farther east are the constellations of **Apus**, **Chamaeleon** and **Volans**.

NASA's *Europa Clipper*

The key to understanding the origins of life on Earth may lie in our observations of objects in our Solar System where water is present. Astrobiologists believe water is a necessary ingredient for the evolution of life, and as the discovery of water beyond the Earth ignites further investigation, planetary scientists have their eyes on **Europa**, a moon of **Jupiter**. NASA's *Europa Clipper* was launched on 14 October 2024 and is due to arrive in April 2030, after traversing a distance of 2.9 billion km. The spacecraft will conduct 49 close flybys of Europa, reaching within 25 km of the icy surface. There are nine instruments on board, and it will be powered by the Sun via large solar arrays, necessary due to the vast distance from the Sun.

Europa is the smallest of the Galilean moons and the sixth largest in the Solar System, it is a quarter of the size of the Earth – 67 Europas would fit inside our planet. It orbits Jupiter once every 3.6 days at a distance almost double that of the Moon from the Earth. Its surface gravity is 13 per cent of the strength of Earth's and one hemisphere is continuously facing Jupiter, so from the near-facing surface one would see Jupiter permanently overhead in the sky. Europa was studied by **NASA's** *Galileo* mission in the late 1990s, during which the spacecraft made 11 flybys. It was also imaged by NASA's *Juno* spacecraft in October 2023 and it will be visited by ESA's *Jupiter Icy Moon Explorer* (*JUICE*), launched 14 April 2023. *JUICE* is currently on its way to the largest of Jupiter's moons, **Ganymede**, and it will conduct two flybys of Europa as part of its mission.

In 2014 and 2016 the **Hubble Space Telescope** imaged plumes erupting from the surface of the moon in ultraviolet light; the Galileo mission revealed distortions in Jupiter's magnetic field around Europa. This is evidence of a magnetic field from Europa interacting with the Jovian field, possibly induced by a deep layer of electrically conductive fluid such as saltwater. A radar on the *Clipper* will penetrate the surface to look for water; the magnetic and gravitational fields of the moon will be measured to further probe the internal structure. The moon is surrounded by a thin oxygen atmosphere; the pressure is barely 100-billionth of Earth's. The *Clipper* will study the plumes and how the atmosphere interacts with Jupiter, and high-resolution images and spectra revealing the chemical and physical properties of the atmosphere will be sent to the NASA team during the mission.

The surface of Europa is relatively smooth and there are few craters. It is estimated the surface is quite young – 40–90 million years old – a result of continual geological activity. Tidal (gravitational) forces from Jupiter have resulted in flexing of the surface and the production of numerous linear cracks, ridges, pits and domes. The water ice shell is estimated to be 15–25 km thick, and is lying on top of an ocean of salty water 60–150 km deep. If confirmed by the *Clipper*, this means Europa contains more than twice as much water as all of Earth's oceans combined. Scientists suggest there could be hydrothermal vents at the boundary between the ocean and rock interior close to the metallic core, another potential site for primitive organisms.

Europa, imaged by NASA's Juno *spacecraft, 2022.*

O

November

November – Introduction

Meteors

Two meteor showers begin in September or October and continue into November. The **Orionids**, one of the streams associated with Comet 1/P Halley, continue until about 7 November. The **Southern Taurids** begin on 10 September and continue until 20 November. The **Northern Taurid** shower, which began in mid-October, reaches maximum – although only with a rate of about five meteors per hour – on 12 November. The Moon is waxing crescent and setting relatively early; the meteors are best observed later in the evening. The shower gradually trails off, ending around 10 December.

Because of the location of the radiant, the **Leonid** shower is best seen from the northern hemisphere, but southern observers may see some rising from the horizon. They have a relatively short period of activity (6 to 30 November), with maximum in the early hours of 18 November. This shower is associated with Comet 55P/Tempel-Tuttle and has shown extraordinary activity on various occasions with many thousands of meteors per hour. The rate in 2026 is likely to be about 10–15 per hour at peak activity; however, the Moon will be waxing gibbous. Conditions will be favourable after midnight. These meteors are the fastest shower meteors recorded (about 70 km per second) and often leave persistent trains.

There is a minor southern meteor shower called the **Phoenicids** that begins activity in late November. Little is known of the shower, partly because the parent comet is believed to be the disintegrated comet D/1819 W1 (Blanpain). With no accurate knowledge of the location of the remnants of the comet, predicting the possible rate becomes little more than guesswork, but the rate is variable and may rapidly increase (as might be expected) if the orbit is nearby. Bright meteors tend to be quite frequent and the meteors are fairly slow. The radiant is located within **Phoenix**, not far from the border with **Eridanus** and the bright star **Achernar** (α Eridani).

The planets

Mercury is placed in **Libra** close to the Sun at sunset. It leads the Sun in the west after 4 November and moves into **Virgo** and then back into **Libra** as the month progresses (mag. 3.3 to 6.5, brightening to −0.7). On 20 November Mercury is at western elongation (mag. −0.5). **Venus** is visible at dawn in Virgo, receding from the Sun in the sky (mag. −4.2 to −4.9). On 10 November it lies 0.1°S of **Spica** (mag. −4.7). **Mars** is in **Leo** and creeps above the horizon around an hour before midnight (mag. 0.8 to 0.4). Mars (mag. 0.7) is 1.2°N of **Jupiter** (mag. −2.1) on 16 November. On 25 November Mars (mag. 0.6) is 1.6°N of **Regulus**. Jupiter is visible in Leo from midnight (mag. −2.0 to −2.2). **Saturn** is in **Cetus** near **Pisces**, in the sky just after sunset (mag. 0.5 to 0.6). **Uranus** is in **Taurus** from early evening (mag. 5.6); it reaches opposition on 25 November at a distance of 18.4 AU from Earth. **Neptune** is in Pisces (mag. 7.7).

N

Sunrise and sunset

City	Date	Sunrise	Sunset
Buenos Aires, Argentina			
	Nov. 01	08:52	22:23
	Nov. 30	08:34	22:51
Cape Town, South Africa			
	Nov. 01	03:46	17:15
	Nov. 30	03:29	17:42
London, UK			
	Nov. 01	06:54	16:34
	Nov. 30	07:42	15:56
Los Angeles, USA			
	Nov. 01	14:13	01:00
	Nov. 30	14:40	00:44
Nairobi, Kenya			
	Nov. 01	03:12	15:21
	Nov. 30	03:16	15:27
Sydney, Australia			
	Nov. 01	18:55	08:23
	Nov. 30	18:37	08:50
Tokyo, Japan			
	Nov. 01	21:02	07:46
	Nov. 30	21:31	07:28
Washington, DC, USA			
	Nov. 01	11:35	22:07
	Nov. 30	12:07	21:47
Wellington, New Zealand			
	Nov. 01	17:08	07:01
	Nov. 30	16:43	07:36

NB: the times given are in Universal Time (UT)

The Moon's phases and ages

Northern Hemisphere

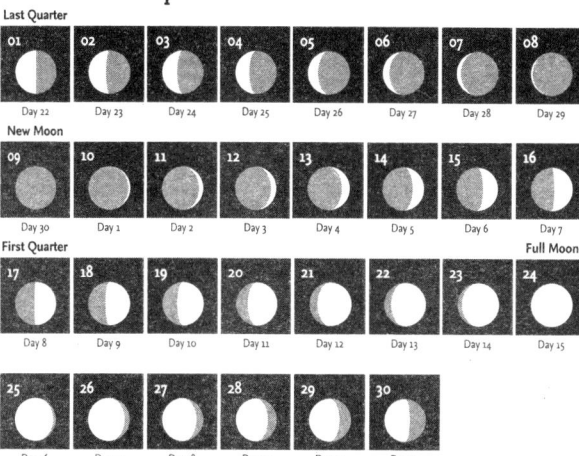

Southern Hemisphere

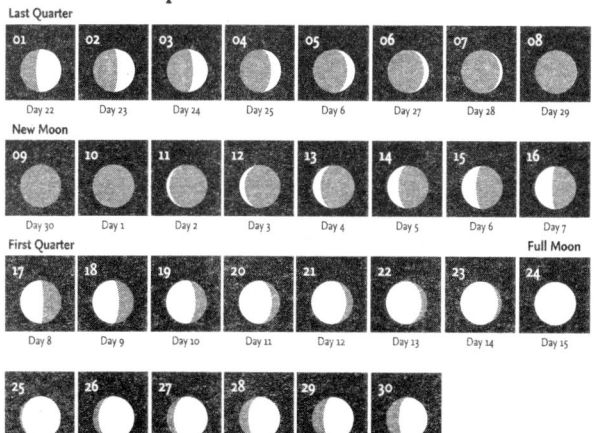

The Moon

The Moon in November

On 1 November the Moon is Last Quarter, a day later the waning gibbous Moon is 1.1°S of **Mars** (mag. 0.8) and 0.5°S of **Jupiter** (mag. –2.1). On 3 November **Regulus** is 0.8°N of the waning crescent Moon. On 7 November the Moon joins **Venus** (mag. –4.5) in **Virgo**, sitting 1.1°S of the planet. It is also 2.4°S of **Spica**. On 9 November the Moon is New and two days later the slender waxing crescent Moon moves 0.3°S of **Antares**. A First Quarter Moon dominates the sky on 17 November; a week later on 24 November the Full Moon is 0.9°N of the **Pleiades**. On 28 November the waning gibbous Moon lies 4.2°S of **Pollux**; two days later it is with Jupiter (mag. –2.2), 1.2°S of the planet and 1.1°S of **Regulus** in **Leo**. On the same day Mars (mag. 0.4) is 3.3°N of the Moon.

Follow the water: Uranus and its moons

Recent new analysis of measurements taken by the **Voyager 2** spacecraft in 1986 suggest that five of the moons of **Uranus** could have hidden water lying beneath their surfaces. As the spacecraft flew past the moons, it may have encountered a powerful solar wind that temporarily distorted the magnetic field around Uranus, thus impacting the measurements taken. After recent investigation reported in *Nature* in November 2024, our understanding of these moons has changed dramatically. Scientists are particularly excited about the moon **Miranda**, one-seventh of the size of the Moon. Scientists now believe a vast ocean was present 100–500 million years ago, and it may still be there today. The results will impact the development of NASA's future mission to Uranus.

Life on the Moon

A mission to grow plants on the Moon is scheduled to launch in 2025, with the seeds being carried in a lunar lander. The **Aleph** mission (Australian Lunar Experiment Promoting Horticulture) is a collaborative project, with the goal of producing food for future astronauts to eat when they are based on the Moon. If the food production is successful, the project will extend to **Mars**.

Miranda, imaged by NASA's Voyager 2 *spacecraft, 1986.*

The plants will need to survive extreme conditions in low gravity, as lunar gravity is one-sixth of the Earth's. Temperatures range from 130°C to below zero. The plants are already adapted to desert conditions and can be dormant during periods of drought, resurrecting when water is present.

NASA's *Artemis* mission to send people back to the Moon is underway. The goal is to create human habitation stations on the lunar surface, which will eventually be a springboard for excursions to Mars. The first stage of the mission, the launch of the uncrewed *Orion* spacecraft, took place on 16 November 2022, after a six-year delay. Stage two is due to take place in 2026, when a crew of four men and women will launch to the Moon, fly past and return to Earth. The third stage, when a crew will land on the Moon, is scheduled for 2027, and there are at least three more stages planned up to 2031 and beyond.

Calendar for November

01	20:28	Last Quarter Moon
02	14:23	Mars (mag. 0.8) 1.1°N of the Moon. Occultation visible from southern French Polynesia, Cook Islands and Pitcairn in the South Pacific Ocean.
02	23:11	Jupiter (mag. −2.1) 0.5°N of the Moon. Occultation visible from Australia, Indonesia, southern India and Malaysia.
03	08:40	Regulus 0.8°N of the Moon. Occultation visible from Brazil, Argentina, Peru and Bolivia.
07	11:31	Venus (mag. −4.5) 1.1°N of the Moon. Occultation visible from Antarctica, Argentina, Chile and Falkland Islands.
07	12:40	Spica 2.4°N of the Moon
09	07:02	New Moon
10	13:49	Venus (mag. −4.7) 0.1°S of Spica
11	03:58	Antares 0.3°N of the Moon. Occultation visible from Samoa, Tonga, American Samoa and southwestern Fiji.
12		Northern Taurid meteor shower maximum
13	17:50	Moon at apogee = 405,619 km
16	06:24	Mars (mag. 0.7) 1.2°N of Jupiter
17	11:48	First Quarter Moon
17–18		Leonid meteor shower maximum
20	23:40	Mercury at greatest elongation (19.6°W, mag. −0.5)
24	11:18	Pleiades 0.9°S of the Moon
24	14:53	Full Moon
25	07:47	Mars (mag. 0.6) 1.6°N of Regulus
25	20:58	Moon at perigee = 359,348 km
25	22:41	Uranus (mag. 5.6) at opposition
28	01:27	Pollux 4.2°N of the Moon
30	09:18	Jupiter (mag. −2.2) 1.2°N of the Moon. Occultation visible from southern Argentina, Chile, Antarctica and Falkland Islands.
30	14:35	Regulus 1.1°N of the Moon. Occultation visible from New Zealand and Norfolk Island.
30	19:32	Mars (mag. 0.4) 3.3°N of the Moon

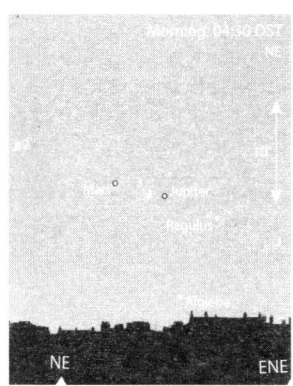

3 November: *The Last Quarter Moon with Mars and Jupiter, Regulus close by (as seen from Sydney).*

7 November: *Venus and the waning crescent Moon in Virgo visible shortly before sunrise. Arcturus visible in the eastern sky (as seen from London).*

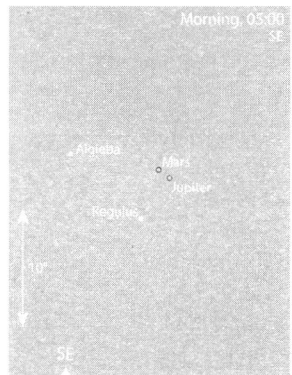

16 November: *Mars and Jupiter in conjunction in Leo, near Regulus in the southeastern sky (as seen from central USA).*

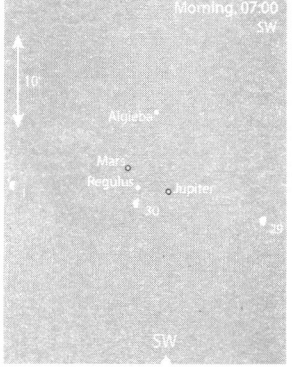

30 November: *Mars, Jupiter and the waning gibbous Moon, near Regulus in Leo (as seen from central USA).*

N

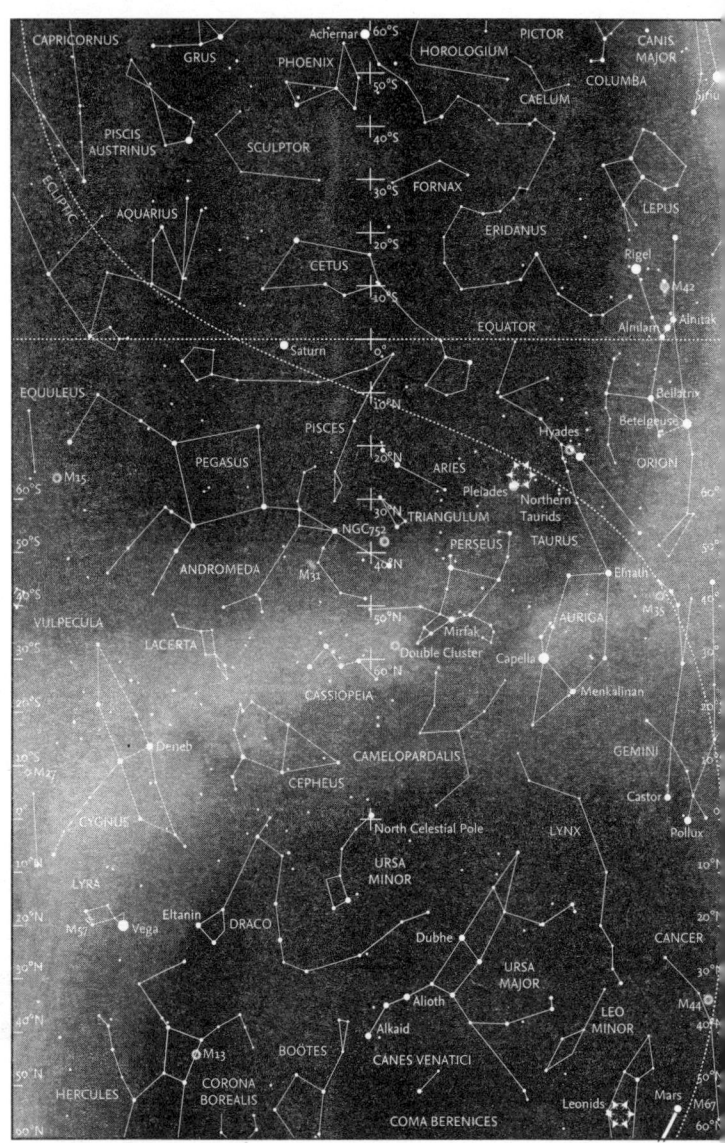

November – Looking North

For observers at mid-northern latitudes, the Milky Way is now high in the north and northwest. Most of **Aquila** has now disappeared below the horizon, but two of the stars of the Summer Triangle, **Vega** in **Lyra** and **Deneb** in **Cygnus**, are still clearly visible in the west. The head of **Draco** is now low in the northwest, **Eltanin** (γ Draconis), the brightest star in the 'head', is at about the same altitude as Vega and only a small portion of **Hercules** remains above the horizon, making the stars difficult to see. **Ursa Minor**, together with **Polaris** itself, is slightly higher in the sky.

The southernmost stars of **Ursa Major** are now coming into view. The Milky Way arches overhead, with the denser star clouds in the west and the less heavily populated region through **Auriga** and **Monoceros** in the east. The star β **Monocerotis** is actually a triple star system; all three stars are 6 to 7 times the mass of the Sun. High overhead, **Cassiopeia** is near the zenith and **Cepheus** has swung round to the northwest, while Auriga is now high in the northeast. **Gemini**, with **Castor** and **Pollux**, is well clear of the eastern horizon, and **Procyon** (α Canis Minoris), in reality a binary star system, is just climbing into view almost due east.

Most of the fainter stars in the constellations to the south and east are easily visible, as is the undistinguished constellation of **Canes Venatici** to the southwest. Another constellation, often ignored, is **Leo Minor** to the southeast of Ursa Major. This consists of little more than three faint stars. Slightly towards the south are the northernmost stars of **Boötes**, although its red giant star **Arcturus** is below the northern horizon.

N

November – Looking South

Several major constellations dominate the southern sky. *Orion* has now risen above the eastern horizon, and part of the long, straggling constellation of *Eridanus* (which begins near *Rigel* (β Orionis)) is visible to the west of Orion. The brightest star, *Achernar* (α Eridani), at the very end of the constellation, is on the horizon for observers at a latitude of 30° north. Immediately north and west of Achernar is *Phoenix* and between that and *Aquarius* lie the two constellations of *Sculptor* and *Piscis Austrinus*, the latter with its solitary bright star, *Fomalhaut*. Below Piscis Austrinus is the constellation of *Grus*.

Higher in the east, *Taurus*, with the *Pleiades* cluster, and orange *Aldebaran* are now easy to observe. To their west, both *Pisces* and *Cetus* are close to the meridian. The famous long-period variable star *Mira* (o Ceti), with a typical range of mag. 3.4 to 9.5, is favourably placed for observation. In the southwest, *Capricornus* has slipped below the horizon, but *Aquarius* remains visible. Even farther west, *Altair* may be seen early in the night, but most of *Aquila* has already disappeared from view. *Delphinus*, together with *Sagitta* and *Vulpecula* in the Milky Way, will soon vanish for another year. Both *Pegasus* and *Andromeda* are easy to see, and one of the lines of stars that make up Andromeda finishes close to the zenith, which is also close to one of the outlying stars of *Perseus*, high in the east.

In the southern hemisphere *Crux* and the two brightest stars of *Centaurus*, *Rigil Kentaurus* (α Centauri) and *Hadar* (β Centauri), are extremely low on the southern horizon. The *False Cross* on the *Carina/Vela* border is now higher, and the constellation of *Puppis* as well as *Canopus* (α Carinae), and both the *Large Magellenic Cloud* (LMC) and the *Small Magellenic Cloud* (SMC), are clearly visible. *Pavo* is descending in the southwest and, in the west, most of *Sagittarius* is below the horizon, with Capricornus and *Scorpius* descending behind it. *Corona Australis* is still just visible. In the east, *Canis Major* is now clearly seen, together with the small constellations of *Columba* and *Lepus* above it.

N

Bertil Lindblad: The Milky Way galaxy is a spiral

In April 1920, American astrophysicists held a Great Debate in Washington DC. Harlow Shapley believed that spiral nebulae (galaxies) were situated inside the Milky Way, while Heber Curtis argued that they were 'island universes' outside our galaxy, and were similar in size to the Milky Way. In 1926, Swedish astronomer Bertil Lindblad (1895–1965) proposed that the Milky Way was a rotating spiral galaxy, a distinct, vast collection of stars separate to other galaxies. He built on the theories of Dutch astronomer Jacobus C. Kapteyn who suggested stars moved in groups, going one way or another; Kapteyn called this motion 'star streaming'.

Lindblad measured the absolute magnitudes of stars (their true brightness after taking into account their distance from Earth) and he looked at the trajectories of globular clusters within our galaxy. He realized that stars ahead of us were nearer to the centre and moving faster. Stars further from the centre appeared to move in the opposite direction because they were orbiting at slower speeds. He found that the stars were orbiting the galactic centre in the constellation Sagittarius. His theory was confirmed in 1927 by Jan Oort, one of Kapteyn's students. The Milky Way takes 225 million years to complete one rotation. Oort's observations revealed the galaxy to be moving faster than expected, and from this he deduced the existence of an unidentified mass capable of causing this surge of speed, which was later called 'dark matter'. This is now thought to make up around 90 per cent of the total mass of the Milky Way.

Milky Way over the Atacama Desert, Chile.

December

December – Introduction

The winter solstice on 21 December brings the shortest period of daylight for observers in the northern hemisphere. In the southern hemisphere, this day marks the summer solstice, during which observers experience the longest period of daylight (midsummer's day).

Meteors

There is one significant meteor shower in December (the last major shower of the year). This is the **Geminid** shower, which is visible over the period 4 to 17 December and comes to maximum on 14 December, when the Moon is a waxing crescent; favourable conditions arise later in the evening. It is one of the most active showers of the year, with a peak rate of around 100 meteors per hour. It is the one major shower that shows good activity before midnight. The source of the Geminids is debris from an asteroid called 3200 Phaethon; most meteor showers originate from cometary debris. The Geminids are assumed to consist of denser, rocky material, they are slower than most other meteors and often appear to last longer. The brightest often break up into numerous luminous fragments that follow similar paths across the sky. There is a second shower: the **Ursids**, active in the last half of the month and peaking on the 22nd, with a rate at maximum of 5 to 10, occasionally rising to 25 per hour. The peak in 2026 occurs when the Moon is a waxing gibbous rising late in the evening, so conditions are unfavourable. The parent body is Comet 8P/Tuttle.

The **Phoenicid** shower visible in the southern hemisphere continues into December, reaching its weak maximum on

The European Space Agency launched its **Proba-3** mission on 4 December 2024. Two spacecraft will line up with the Sun in an arrangement that will stay precisely aligned within 1 mm to form an artificial solar eclipse. The lead spacecraft (called the Occular satellite) carries a 140-cm wide disc that blocks out the Sun. It casts a shadow on the second spacecraft (called the Coronagraph satellite), which sits 144 m behind the first. This satellite will observe the faint halo or corona around the Sun, only visible during a solar eclipse.

2 December. The ***Puppid-Velid*** shower's radiant is on the border between the two constellations. The shower begins on 1 December, lasting until 15 December, with maximum on 7 December. It is a weak shower with a maximum hourly rate of about 10 meteors, but bright meteors are often seen.

The planets

Mercury sits in ***Libra*** then passes through ***Scorpius*** and ***Ophiuchus*** and ***Sagittarius***; it approaches the Sun, switching from appearing at dawn to dusk after New Year's Day (mag. −0.7 to −1.3). ***Venus*** is in ***Virgo***, visible at dawn; it passes into Libra and reaches western elongation in the new year on 3 January 2027 (mag. −4.9 to −4.6). ***Mars*** stays in ***Leo*** for the month; it is above the horizon a few hours before midnight by New Year (mag. 0.4 to −0.1). ***Jupiter*** is in Leo and visible later in the night with Mars (mag. −2.2 to −2.4). On 12 December Jupiter (mag. −2.3) is 1.3°N of ***Regulus***. It begins to move westwards in Leo on 13 December. ***Saturn*** is on the border between ***Cetus*** and ***Pisces*** and visible from sunset to just before midnight (mag. 0.6 to 0.7). On 11 December ***Saturn*** ends retrograde motion at mag. 0.6. ***Uranus*** is settled in ***Taurus*** and visible all night (mag. 5.6). ***Neptune*** continues in Pisces at mag. 7.7 to 7.8; it ends retrograde motion on 12 December, moving eastwards towards the end of the year.

D

Sunrise and sunset

City	Date	Sunrise	Sunset
Buenos Aires, Argentina			
	Dec. 01	08:34	22:52
	Dec. 31	08:44	23:10
Cape Town, South Africa			
	Dec. 01	03:29	17:43
	Dec. 31	03:38	18:01
London, UK			
	Dec. 01	07:44	15:55
	Dec. 31	08:06	16:01
Los Angeles, USA			
	Dec. 01	14:40	00:44
	Dec. 31	14:58	00:54
Nairobi, Kenya			
	Dec. 01	03:16	15:27
	Dec. 31	03:30	15:42
Sydney, Australia			
	Dec. 01	18:37	08:51
	Dec. 31	18:47	09:09
Tokyo, Japan			
	Dec. 01	21:32	07:28
	Dec. 31	21:50	07:37
Washington, DC, USA			
	Dec. 01	12:08	21:46
	Dec. 31	12:27	21:56
Wellington, New Zealand			
	Dec. 01	16:43	07:37
	Dec. 31	16:50	07:57

NB: the times given are in Universal Time (UT)

The Moon's phases and ages

Northern Hemisphere

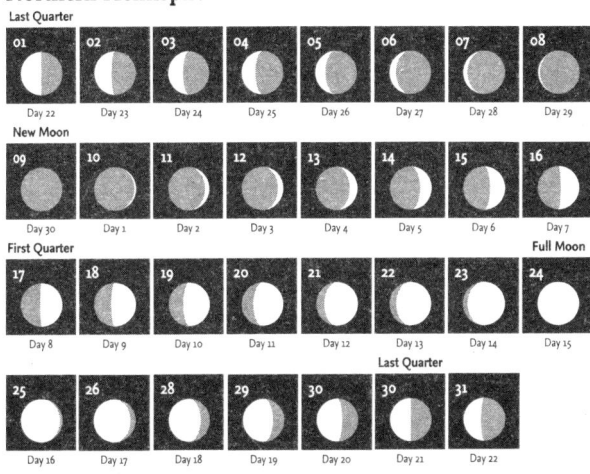

Last Quarter

01 Day 22	02 Day 23	03 Day 24	04 Day 25	05 Day 26	06 Day 27	07 Day 28	08 Day 29

New Moon

09 Day 30	10 Day 1	11 Day 2	12 Day 3	13 Day 4	14 Day 5	15 Day 6	16 Day 7

First Quarter   Full Moon

17 Day 8	18 Day 9	19 Day 10	20 Day 11	21 Day 12	22 Day 13	23 Day 14	24 Day 15

Last Quarter

25 Day 16	26 Day 17	28 Day 18	29 Day 19	30 Day 20	30 Day 21	31 Day 22

Southern Hemisphere

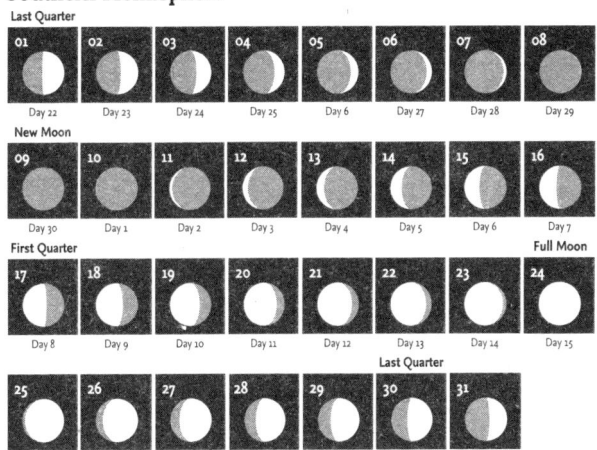

Last Quarter

01 Day 22	02 Day 23	03 Day 24	04 Day 25	05 Day 6	06 Day 27	07 Day 28	08 Day 29

New Moon

09 Day 30	10 Day 1	11 Day 2	12 Day 3	13 Day 4	14 Day 5	15 Day 6	16 Day 7

First Quarter   Full Moon

17 Day 8	18 Day 9	19 Day 10	20 Day 11	21 Day 12	22 Day 13	23 Day 14	24 Day 15

Last Quarter

25 Day 16	26 Day 17	27 Day 18	28 Day 19	29 Day 20	30 Day 21	31 Day 22

D

227

The Moon

The Moon in December

On 1 December the Moon is Last Quarter; as it wanes it moves to 2.5°S of *Spica*. A New Moon arises on 9 December, two days later it is at its farthest point from the Earth (apogee), a distance of 406,421 km. First Quarter is on 17 December. The waxing gibbous Moon is 1.0°N of the **Pleiades** on 21 December; three days later it is Full. On 24 December the Moon makes its closest approach to Earth, with a perigee distance of 356,650 km. On 25 December the waning gibbous Moon is 4.4°S of **Pollux**; two days later it moves to 1.5°S of **Jupiter** (mag. −2.4) and 1.4°S of **Regulus** in **Leo**. The Last Quarter Moon occurs on 30 December.

Venus was never habitable

A study of the atmosphere of **Venus** has revealed the internal structure is dry – it never contained oceans of liquid water and is therefore very unlikely to have harboured life. Venus has an average temperature of 460°C, higher than that of the closest planet to the Sun – **Mercury**. The Venusian atmosphere is thick with carbon dioxide and sulfuric acid clouds, and the pressure is 90 times that of Earth's at sea level. It used to be believed by planetary scientists that Venus may have had water on its surface until a runaway greenhouse effect drastically changed its climate. Researchers at the University of Cambridge published their analysis of the composition of the atmosphere in *Nature* in December 2024. They found that no water was present in volcanic eruptions and therefore concluded there is no water beneath the surface.

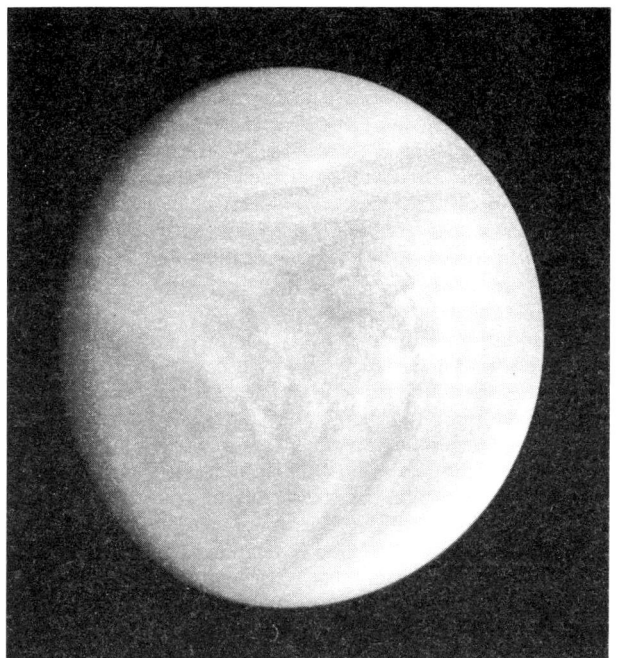

Ultraviolet image of the clouds of Venus.

On Christmas Eve 2024, NASA's **Parker Solar Probe** passed 6.1 million km from the surface of the Sun during its closest approach (perihelion), the closest a spacecraft has ever reached. It is in a highly eccentric elliptical orbit, looping the Sun once every 88 days. At perihelion the probe was exposed to a temperature of 980°C, flying though the low-density but very hot corona – a region encompassing the Sun and extending out into space to a distance of millions of kilometres. The probe survived the encounter, sending telemetry at the beginning of 2025.

D

The supermassive black hole in the Milky Way

In 1933, Karl Jansky found a strong source of radio waves in the direction of the galactic centre in Sagittarius. In 1974, radio telescopes provided more detailed images. The strongest source was called **Sagittarius A***, and in the 1980s a supermassive black hole was proposed as the cause of the emission. In 2002, German astrophysicist Reinhard Genzel at the Max Planck Institute for Extraterrestrial Physics near Munich reported the orbital path of a star called S2 around the galactic centre; his results strengthened the case for the presence of a black hole.

In December 2008, American astrophysicist Andrea Ghez and her team at the University of California, Los Angeles, used the orbital motions of the stars to determine the mass of Sagittarius A*; the current best estimate is 4.297 million solar masses. She was awarded the Nobel Prize in Physics in 2020, along with Reinhard Genzel, for their discovery of the supermassive black hole lurking at the heart of our galaxy. The first image of Sagittarius A* was released on 12 May 2022, taken by the Event Horizon Telescope, an array of radio dishes around the world used to take high-resolution images.

Data from NASA's *Dawn* mission reveals the existence of organic materials on the dwarf planet **Ceres**. Ceres is the largest object in the asteroid belt, 3.7 times smaller in diameter than the Moon. Water ice is present on its surface; it is thought there may be extensive liquid water beneath the surface – the organic compounds are thought to exist near large craters and originated from deep within Ceres.

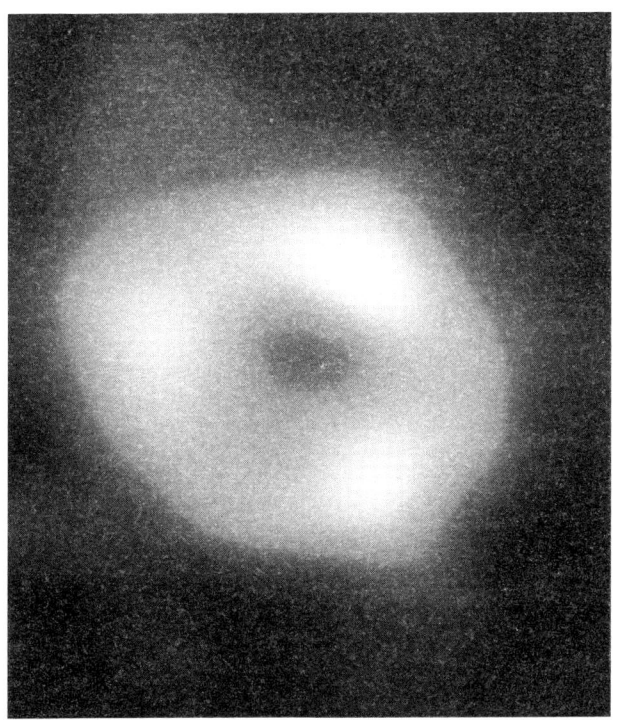

Radio image of Sagittarius A*, the supermassive black hole at the centre of the Milky Way. Event Horizon Telescope, 2022.

D

Calendar for December

01	06:09	Last Quarter Moon
02		Phoenicid meteor shower maximum
04	18:36	Spica 2.5°N of the Moon
07		Puppid-Velid meteor shower maximum
09	00:52	New Moon
11	06:46	Moon at apogee = 406,421 km
12	15:35	Jupiter (mag. −2.3) 1.3°N of Regulus
14		Geminid meteor shower maximum
17	05:43	First Quarter Moon
21	20:50	Winter Solstice
21	22:37	Pleiades 1.0°S of the Moon
22		Ursid meteor shower maximum
24	01:28	Full Moon
24	08:30	Moon at perigee = 356,650 km
25	11:41	Pollux 4.4°N of the Moon
27	17:32	Jupiter (mag. −2.4) 1.5°N of the Moon
27	22:44	Regulus 1.4°N of the Moon
30	18:59	Last Quarter Moon

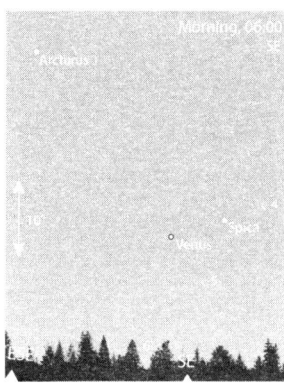

5 December: *Waning crescent Moon close to Venus and Spica, Arcturus visible in the east (as seen from London).*

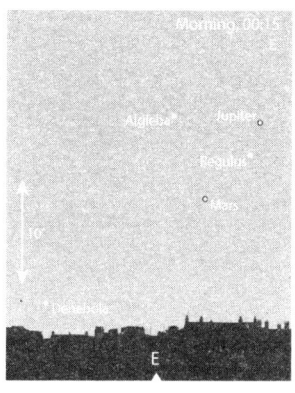

13 December: *Jupiter and Regulus appear together, Mars due east in Leo (as seen from London).*

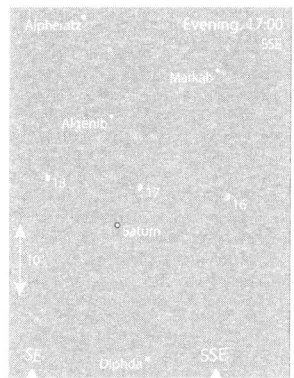

17 December: *The First Quarter Moon close to Saturn in Pisces in the early evening sky. Diphda (β Cet), Algenib (γ Peg) and Markab (α Peg) are visible in the same area of the sky (as seen from central USA).*

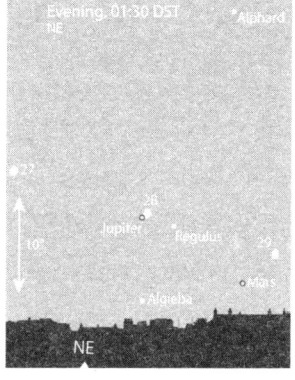

27 December: *The waning gibbous Moon next to Regulus. Jupiter and Mars are close by (as seen from Sydney).*

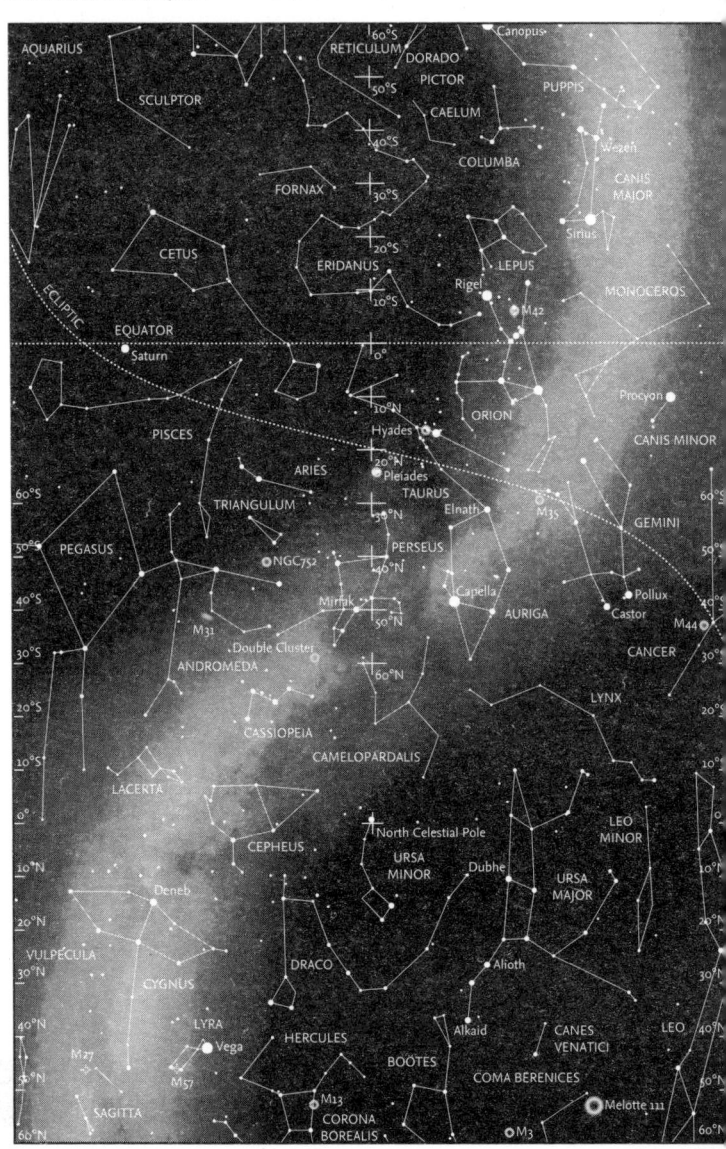

December – Looking North

For observers close to the equator, **Orion** is high overhead, to the east of the meridian, with **Taurus**, **Auriga** (with **Capella**) and **Gemini** way above the northern horizon. For observers at mid-northern latitudes, it is **Perseus** that is at the zenith, with **Andromeda** stretching off to the west and the Great Square of **Pegasus** in the northwest.

Ursa Major has now swung around and is starting to 'climb' in the east. The fainter stars in the southern part of the constellation are now fully in view. The other bear, **Ursa Minor**, 'hangs' below **Polaris** in the north. Below the 'tail' is the inconspicuous constellation of **Canes Venatici**, and some observers at high latitudes may even be able to glimpse some of the stars of **Coma Berenices**, low on the horizon in the northeast. Directly above it is the faint constellation of **Camelopardalis**, with the other circumpolar constellation, **Lynx**, to its east. **Vega** (α Lyrae) is skimming the horizon in the northwest, but **Deneb** (α Cygni) and most of **Cygnus** remain visible farther west. In the east, **Regulus** (α Leonis) and the constellation of **Leo** are beginning to rise above the horizon. **Cancer** stands high in the east, with Gemini even higher in the sky. Even farther east are the two bright stars of Gemini: **Castor** and **Pollux**. December is a good time to examine the star clouds of the fainter portion of the Milky Way, between **Cassiopeia** in the west to Gemini and Orion in the east. Farther down is the zig-zag constellation of **Lacerta**, which is like **Cepheus**, in that both of them are partly within the band of stars. Much of the constellation of **Hercules** is clear of the northern horizon, together with some of the northernmost stars in **Boötes**.

D

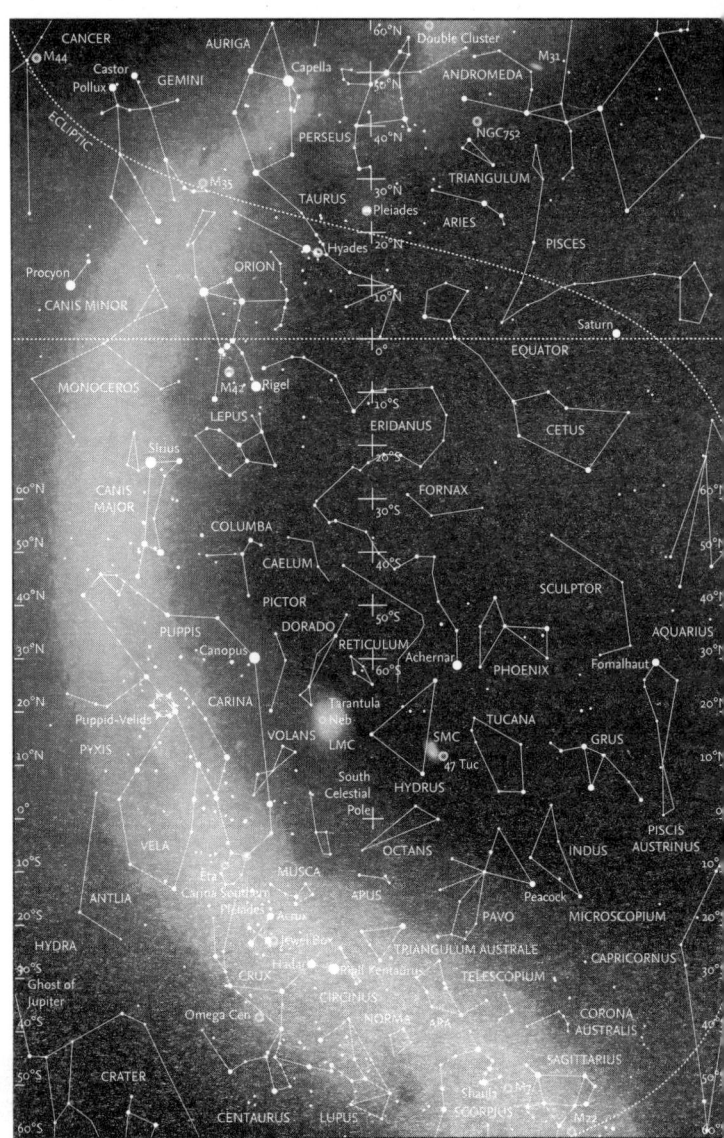

December – Looking South

The fine open cluster of the **Pleiades** is due south late in the night, with the **Hyades** cluster, **Aldebaran** and the rest of **Taurus** clearly visible to the east. **Auriga** (with **Capella**) and **Gemini** (with **Castor** and **Pollux**) are both well-placed for observation. **Orion** has made a welcome return to the winter sky (in the northern hemisphere), and both **Canis Minor** (with **Procyon**) and **Canis Major** (with **Sirius**, the brightest star in the sky) are now well above the horizon. The long, winding constellation of **Eridanus** begins near **Rigel** in Orion and runs south, partly enclosing **Fornax** (which was once part of the larger constellation), until it ends at **Achernar** (α Eridani). Between Eridanus and Canis Major is the tiny, faint constellation of **Caelum** and the larger and brighter **Columba**.

The small, poorly known constellation of **Lepus** lies to the south of Orion. In the west, **Aquarius** has now disappeared, and **Cetus** is becoming lower, but **Pisces** is still easily seen, as are the constellations of **Aries**, **Triangulum** and **Andromeda** above it. The Great Square of **Pegasus** is starting to plunge down towards the western horizon and, because of its orientation on the sky, appears more like a large diamond standing on one point than a square.

Further south, **Crux** and the two brightest stars in **Centaurus**, **Hadar** and **Rigil Kentaurus**, are now higher above the horizon. The **Eta Carina Nebula** and the **Southern Pleiades** are now conveniently placed for observation. Above them, the **False Cross** is clearly seen, with the whole of **Vela** and below it the constellation of **Antlia** (the air pump). **Carina** with **Canopus** (α Carinae) and **Puppis** are roughly halfway between the horizon and the zenith. In the west, Achernar and **Phoenix** are about the same altitude as Canopus. The faint constellations of **Pictor**, **Dorado**, **Reticulum** and **Horologium** lie between them. **Pavo** with **Peacock** (α Pavonis) is becoming low, as are the constellations of **Indus**, **Grus** and **Piscis Austrinus**. Higher still are the constellations of **Sculptor** and, near the zenith, **Fornax**. Those in the far south are able to see all the stars in Centaurus and **Lupus**, together with the southernmost stars of **Scorpius** (the 'sting') and **Sagittarius**, as well as other constellations, such as **Norma**, **Ara**, **Telescopium** and **Corona Australis**.

Additional
Information

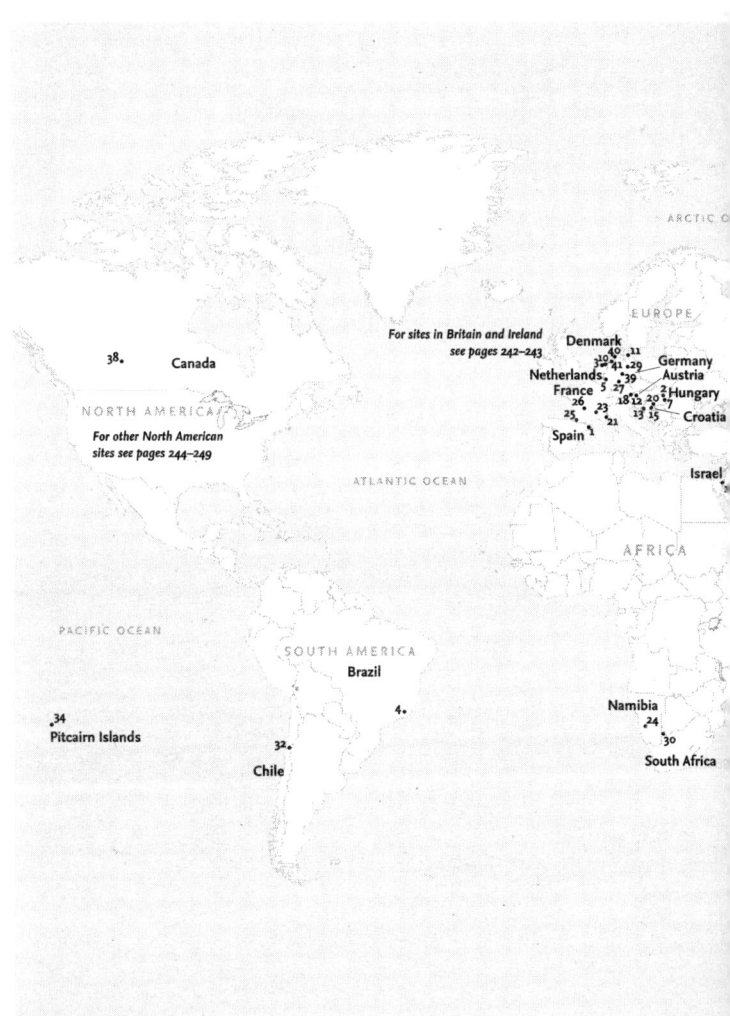

Canada

NORTH AMERICA

38.

For other North American sites see pages 244–249

For sites in Britain and Ireland see pages 242–243

EUROPE

Denmark

Netherlands

France

Spain

Germany
Austria
Hungary
Croatia

Israel

ATLANTIC OCEAN

AFRICA

PACIFIC OCEAN

SOUTH AMERICA

Brazil

Chile

Pitcairn Islands

Namibia

South Africa

ARCTIC O

ARCTIC OCEAN

EUROPE

Germany
Austria
2 Hungary
20 7
15 Croatia

ASIA

Israel
14

South 19 37 9
Korea
Japan
6 8
Taiwan

AFRICA

PACIFIC OCEAN

INDIAN OCEAN

OCEANIA

33
Niue

ia
24
30
South Africa

Australia 36

28 17

31

New Zealand

22 16

35

Dark-Sky Sites, UK & Ireland

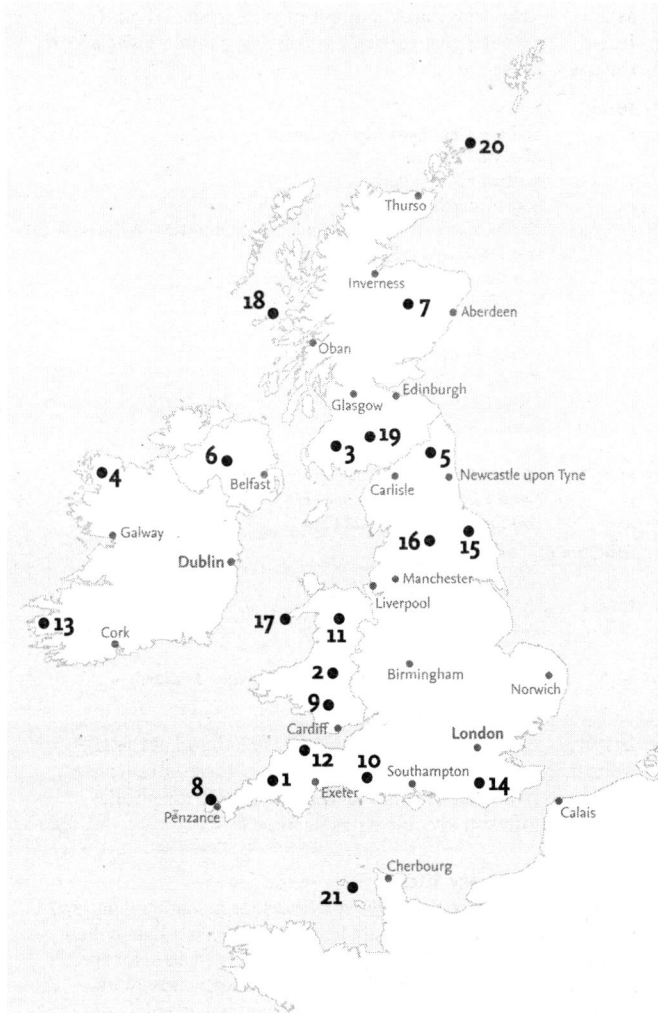

The **International Dark-Sky Association** (IDA) recognizes various categories of sites that offer areas where the sky is dark at night, free from light pollution and particularly suitable for astronomical observing. A number of sites in the UK and Ireland have been given specific recognition and are shown on the map. These are:

Parks

1	*Bodmin Moor Dark Sky Landscape*
2	*Elan Valley Estate*
3	*Galloway Forest Park*
4	*Mayo Dark Sky Park*
5	*Northumberland National Park and Kielder Water & Forest Park*
6	*OM Dark Sky Park & Observatory*
7	*Tomintoul and Glenlivet – Cairngorms*
8	*West Penwith*

Reserves

9	*Brecon Beacons National Park*
10	*Cranborne Chase*
11	*Eryri National Park (Snowdonia)*
12	*Exmoor National Park*
13	*Kerry*
14	*Moore's Reserve South Downs National Park*
15	*North York Moors National Park*
16	*Yorkshire Dales National Park*

Sanctuaries

17	*Ynys Enlli (Bardsey Island)*

Communities

18	*Coll (Inner Hebrides, Scotland)*
19	*Moffat*
20	*North Ronaldsay Dark Sky Island (Orkney, Scotland)*
21	*Sark (Channel Islands)*

Details of these sites and web links may be found at the IDA website: https://www.darksky.org/ Many of these sites have major observatories or other facilities available for public observing (often at specific dates or times).

Dark Sky Discovery Sites

In the UK there is also the **Dark Sky Discovery** organization. This gives recognition to smaller sites, again free from immediate light pollution, that are open to observing at any time. Some sites are used for specific public observing sessions. A full listing of sites is at: https://www.darkskydiscovery.org.uk/ but specific events are publicized locally.

US Dark-Sky Sites

Dark Sky Discovery Sites

DarkSky International, formerly the International Dark-Sky Association (IDA), recognizes various categories of sites that offer areas where the sky is dark at night, free from light pollution and particularly suitable for astronomical observing. There are numerous sites in the United States, shown on the map and listed here. Further details are at: https://www.darksky.org/.

Information on the various categories and individual sites are at: https://www.darksky.org/our-work/conservation/idsp/. New sites are continually being added, so check the IDA website for details.

Many of these sites have major observatories or other facilities available for public observing (often at specific dates or times).

Parks

1   *AMC Maine Woods* (ME)
2   *Antelope Island State Park* (UT)
3   *Anza-Borrego Desert State Park* (CA)
4   *Arches National Park* (UT)
5   *Big Bend National Park* (TX)
6   *Big Bend Ranch State Park* (TX)
7   *Big Cypress National Preserve* (FL)
8   *Black Canyon of the Gunnison National Park* (CO)
9   *Bryce Canyon National Park* (UT)
10 *Buffalo National River* (AR)
11 *Canyonlands National Park* (UT)
12 *Cape Lookout National Seashore* (NC)
13 *Capitol Reef National Park* (UT)
14 *Capulin Volcano National Monument* (NM)
15 *Cedar Breaks National Monument* (UT)
16 *Chaco Culture National Historical Park* (NM)
17 *Cherry Springs State Park* (PA)
18 *Chiricahua National Monument* (AZ)
19 *Clayton Lake State Park* (NM)
20 *Copper Breaks State Park* (TX)
21 *Craters of the Moon National Monument* (ID)
22 *Curecanti National Recreation Area* (CO)
23 *Dead Horse Point State Park* (UT)
24 *Death Valley National Park* (CA)
25 *Dinosaur National Monument* (CO)
26 *Dr. T.K. Lawless County Park* (MI)
27 *East Canyon State Park* (UT)
28 *El Morro National Monument* (NM)
29 *Enchanted Rock State Natural Area* (TX)
30 *Flagstaff Area National Monuments* (AZ)
31 *Florissant Fossil Beds National Monument* (CO)
32 *Fort Union National Monument* (NM)

66 *Pisgah Astronomical Research Institute* (NC)

67 *Prineville Reservoir State Park* (OR)

68 *Rappahannock County Park* (VA)

69 *Rockport State Park* (UT)

70 *Salinas Pueblo Missions National Monument* (NM)

71 *Sky Meadows State Park* (VA)

72 *South Llano River State Park* (TX)

73 *Staunton River State Park* (VA)

74 *Steinaker State Park* (UT)

75 *Stephen C. Foster State Park* (GA)

76 *Tonto National Monument* (AZ)

77 *Top of the Pines* (CO)

78 *Tumacácori National Historical Park* (AZ)

79 *UBarU Camp and Retreat Center* (TX)

80 *Valles Caldera National Preserve* (NM)

81 *Voyageurs National Park* (MN)

82 *Waterton-Glacier International Peace Park* (Canada/MT)

83 *Watoga State Park* (WV)

84 *Zion National Park* (UT)

Reserves

85 *Central Idaho* (ID)

Sanctuaries

86 *Black Gap Wildlife Management Area* (TX)

87 *Boundary Craters Canoe Area Wilderness* (MN)

88 *Cosmic Campground* (NM)

89 *Devils River State Natural Area – Del Norte Unit* (TX)

90 *Katahdin Woods and Waters National Monument* (ME)

91 *Massacre Rim Wilderness Study Area* (NV)

92 *Medicine Rocks State Park* (MT)

93 *Rainbow Bridge National Monument* (UT)

Canadian Dark-Sky Sites

The Royal Astronomical Society of Canada (RASC) has developed formal guidelines and requirements for three types of light-restricted protected areas: Dark-Sky Preserves, Urban Star Parks and Nocturnal Preserves. The focus of the Canadian Program is primarily to protect the nocturnal environment; therefore, the outdoor lighting requirements are the most stringent, but also the most effective. Canadian Parks and other areas that meet these guidelines and successfully apply for one of these designations are officially recognized. Many parks across Canada have been designated in recent years – see the list below and the RASC website: https://rasc.ca/lpa/dark-sky-sites. New sites are continually being added, so check the RASC website for details.

Dark-Sky Preserves

1   *Au Diable Vert* (QC)

2   *Beaver Hills* (AB)

3   *Bluewater Outdoor Education Centre* (ON)

4   *Bruce Peninsula National Park* (ON)

Dark-Sky Sites Around the World

Parks

1 *Albanyà* (Spain)

2 *Bükk National Park* (Hungary)

3 *De Boschplaat* (Netherlands)

4 *Desengano State Park* (Brazil)

5 *Eifel National Park* (Germany)

6 *Hehuan Mountain* (Taiwan)

7 *Hortobágy National Park* (Hungary)

8 *Iriomote-Ishigaki National Park* (Japan)

9 *Kozushima Dark Sky Island* (Japan) – also a Dark Sky Community

10 *Lauwersmeer National Park* (Netherlands)

11 *Møn and Nyord* (Denmark)

12 *Naturpark Attersee-Traunsee* (Austria)

13 *Petrova Gora-Biljeg* (Croatia)

14 *Ramon Crater Nature Reserve* (Israel)

15 *Vrani kamen* (Croatia)

16 *Wai-Iti* (New Zealand)

17 *Warrumbungle Dark Sky Park* (Australia)

18 *Winklmoosalm* (Germany)

19 *Yeongyang Firefly Eco Park* (South Korea)

20 *Zselic National Landscape Protection Area* (Hungary)

Reserves

21 *Alpes Azur Mercantour* (France)

22 *Aoraki Mackenzie* (New Zealand)

23 *Cévennes National Park* (France)

24 *NamibRand Nature Reserve* (Namibia)

25 *Pic du Midi* (France)

26 *Regional Natural Park of Millevaches in Limousin* (France)

27 *Rhön* (Germany)

28 *River Murray* (Australia)

29 *Westhavelland* (Germany)

Sanctuaries

30 *!Ae!Hai Kalahari Heritage Park* (South Africa)

31 *Aotea / Great Barrier Island* (New Zealand)

32 *Gabriela Mistral* (Chile)

33 *Niue* (New Zealand) – also a Dark Sky Community

34 *Pitcairn Islands* (UK)

35 *Stewart Island / Rakiura* (New Zealand)

36 *The Jump-Up* (Australia)

Communities

37 *Bisei Town, Ibara City* (Japan)

38 *Bon Accord* (Canada)

39 *Fulda, Hesse* (Germany)

40 *Pellworm Star Island* (Germany)

41 *Spiekeroog Star Island* (Germany)

Twilight Diagrams

Sunrise, sunset, twilight

For each individual month, we give details of sunrise and sunset times for nine cities across the world. But observing the stars is also affected by twilight, and this varies considerably from place to place. During the summer, especially at high latitudes, twilight may persist throughout the night and make it difficult to see the faintest stars. Beyond the Arctic and Antarctic Circles, of course, the Sun does not set for 24 hours at least once during the summer (and rise for 24 hours at least once during the winter). Even when the Sun does dip below the horizon at high latitudes, bright twilight persists throughout the night, so observing the stars is impossible.

There are three recognized stages of twilight: civil twilight, when the Sun is less than 6° below the horizon; nautical twilight, when the Sun is between 6° and 12° below the horizon; and astronomical twilight, when the Sun is between 12° and 18° below the horizon. Full darkness occurs only when the Sun is more than 18° below the horizon. During nautical twilight, only the very brightest stars are visible. (These are the stars that were used for navigation, hence the name for this stage.) During astronomical twilight, the faintest stars visible to the naked eye may be seen directly overhead, but are lost at lower altitudes. They become visible only once it is fully dark. The diagrams show the duration of twilight at the various cities. Of the locations shown, during the summer months there is full darkness at most of the cities, but it never occurs during the summer at the latitude of London. Observing conditions are most favourable at somewhere like Nairobi, which is very close to the equator, so there is not only little twilight, and a long period of full darkness, but there are also only slight variations in timing and duration throughout the year.

The diagrams should be used in conjunction with the monthly moon phase diagrams, as moon phase has a large impact on the darkness of the sky.

Buenos Aires, Argentina – Latitude 34.7°S – Longitude 58.5°W

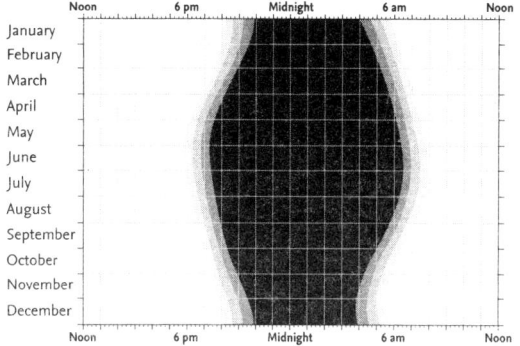

Cape Town, South Africa – Latitude 33.9°S – Longitude 18.5°E

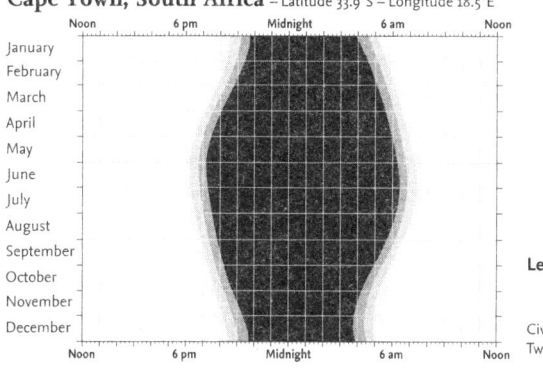

London, UK – Latitude 51.5°N – Longitude 0.2°W

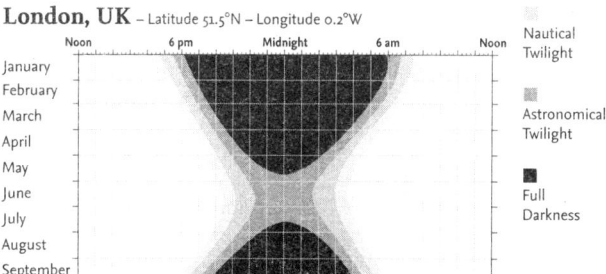

Legend

Civil
Twilight

Nautical
Twilight

Astronomical
Twilight

Full
Darkness

Los Angeles, USA – Latitude 34.0°N – Longitude 118.2° W

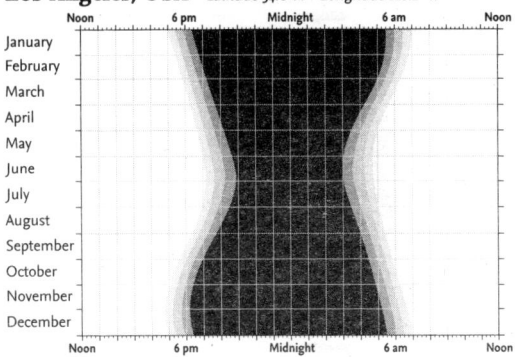

Nairobi, Kenya – Latitude 1.3°S – Longitude 36.8°E

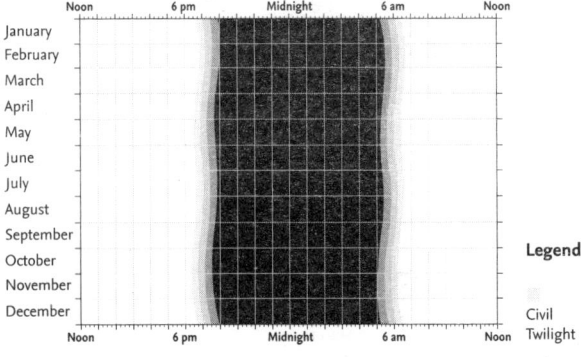

Legend

Civil Twilight

Nautical Twilight

Astronomical Twilight

Full Darkness

Sydney, Australia – Latitude 33.5°S – Longitude 151.2°E

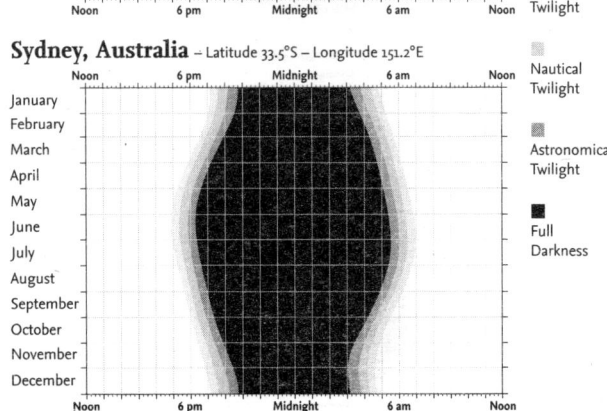

Tokyo, Japan – Latitude 35.7°N – Longitude 139.8°E

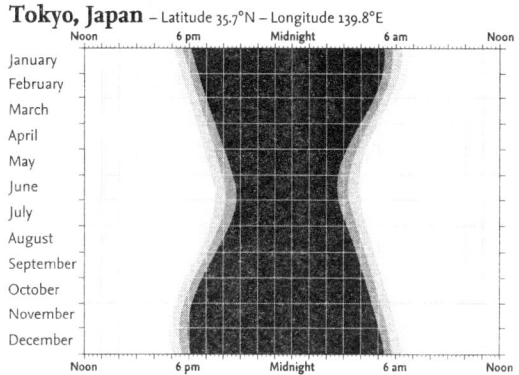

Noon
6 pm
Midnight
6 am
Noon
January
February
March
April
May
June
July
August
September
October
November
December
Noon
6 pm
Midnight
6 am
Noon

Washington, DC, USA – Latitude 38.9°N – Longitude 77.0°W

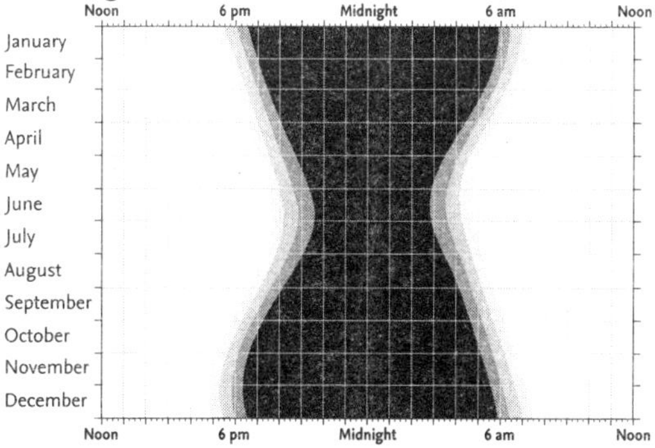

Noon
6 pm
Midnight
6 am
Noon
January
February
March
April
May
June
July
August
September
October
November
December
Noon
6 pm
Midnight
6 am
Noon

Wellington, New Zealand – Latitude 41.3°S – Longitude 174.8°E

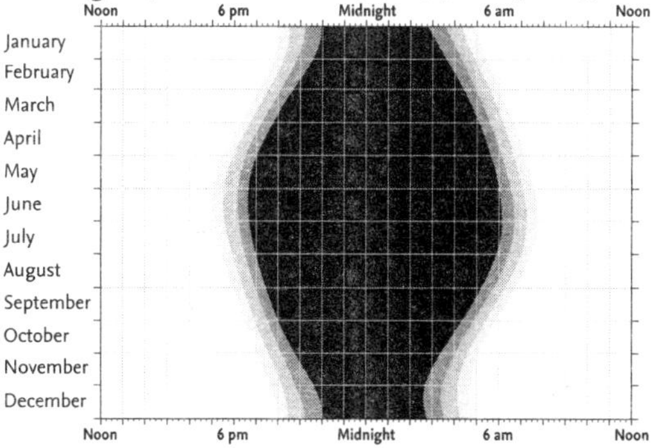

Noon
6 pm
Midnight
6 am
Noon
January
February
March
April
May
June
July
August
September
October
November
December
Noon
6 pm
Midnight
6 am
Noon

Legend

Civil Twilight

Nautical Twilight

Astronomical Twilight

Full Darkness

Glossary and Tables

aphelion	The point on an orbit that is farthest from the Sun.
apogee	The point on its orbit at which the Moon is farthest from the Earth.
appulse	The apparently close approach of two celestial objects in the sky: two planets, or a planet and star.
astronomical unit	(AU) The mean distance of the Earth from the Sun, 149,597,870 km.
celestial equator	The great circle on the celestial sphere that is in the same plane as the Earth's equator.
celestial sphere	The apparent sphere surrounding the Earth on which all celestial bodies (stars, planets, etc.) seem to be located.
conjunction	The point in time when two celestial objects have the same celestial longitude.
	In the case of the Sun and a planet, superior conjunction occurs when the planet lies on the far side of the Sun (as seen from Earth). For Mercury and Venus, inferior conjunction occurs when they pass between the Sun and the Earth.
direct motion	Motion from west to east on the sky.
ecliptic	The apparent path of the Sun across the sky throughout the year.
elongation	The point at which an inferior planet has the greatest angular distance from the Sun, as seen from Earth.
equinox	The two points during the year when night and day have equal duration.
	Also: the points on the sky at which the ecliptic intersects the celestial equator. The vernal (spring) equinox is of particular importance in astronomy.
gibbous	The stage in the sequence of phases at which the illumination of a body lies between half and full. In the case of the Moon, the term is applied to phases between First Quarter and Full, and between Full and Last Quarter.

inferior planet	Either of the planets Mercury or Venus, which have orbits inside that of the Earth.
magnitude	A number to represent brightness of a star, planet or other celestial body. It is a logarithmic scale, where larger numbers indicate fainter brightness. A difference of 5 in magnitude indicates a difference of 100 in actual brightness, thus a first-magnitude star is 100 times as bright as one of sixth magnitude.
meridian	The great circle passing through the North and South Poles of a body and the observer's position; or the corresponding great circle on the celestial sphere that passes through the North and South Celestial Poles and also through the observer's zenith.
nadir	The point on the celestial sphere directly beneath the observer's feet, opposite the zenith.
occultation	The disappearance of one celestial body behind another, such as when stars or planets are hidden behind the Moon.
opposition	The point on a superior planet's orbit at which it is directly opposite the Sun in the sky.
perigee	The point on its orbit at which the Moon is closest to the Earth.
perihelion	The point on an orbit that is closest to the Sun.
retrograde motion	Motion from east to west on the sky.
superior planet	A planet that has an orbit outside that of the Earth.
vernal equinox	The point at which the Sun, in its apparent motion along the ecliptic, crosses the celestial equator from south to north. Also known as the First Point of Aries.
zenith	The point directly above the observer's head.
zodiac	A band, stretching 8° on either side of the ecliptic, within which the Moon and planets appear to move. It consists of twelve equal areas, originally named after the constellation that once lay within it.

The Constellations

There are 88 constellations covering the whole of the celestial sphere. The names themselves are expressed in Latin, and the names of stars are frequently given by Greek letters followed by the genitive of the constellation name. The genitives and English names of the various constellations are included.

Name	Genitive	Abbr.	English name
Andromeda	Andromedae	And	Andromeda
Antlia	Antliae	Ant	Air Pump
Apus	Apodis	Aps	Bird of Paradise
Aquarius	Aquarii	Aqr	Water Bearer
Aquila	Aquilae	Aql	Eagle
Ara	Arae	Ara	Altar
Aries	Arietis	Ari	Ram
Auriga	Aurigae	Aur	Charioteer
Boötes	Boötis	Boo	Herdsman
Caelum	Caeli	Cae	Burin
Camelopardalis	Camelopardalis	Cam	Giraffe
Cancer	Cancri	Cnc	Crab
Canes Venatici	Canum Venaticorum	CVn	Hunting Dogs
Canis Major	Canis Majoris	CMa	Big Dog
Canis Minor	Canis Minoris	CMi	Little Dog
Capricornus	Capricorni	Cap	Sea Goat
Carina	Carinae	Car	Keel
Cassiopeia	Cassiopeiae	Cas	Cassiopeia
Centaurus	Centauri	Cen	Centaur
Cepheus	Cephei	Cep	Cepheus
Cetus	Ceti	Cet	Whale
Chamaeleon	Chamaeleontis	Cha	Chameleon
Circinus	Circini	Cir	Compasses
Columba	Columbae	Col	Dove
Coma Berenices	Comae Berenices	Com	Berenice's Hair
Corona Australis	Coronae Australis	CrA	Southern Crown
Corona Borealis	Coronae Borealis	CrB	Northern Crown

Name	Genitive	Abbr.	English name
Corvus	Corvi	Crv	Crow
Crater	Crateris	Crt	Cup
Crux	Crucis	Cru	Southern Cross
Cygnus	Cygni	Cyg	Swan
Delphinus	Delphini	Del	Dolphin
Dorado	Doradus	Dor	Dorado
Draco	Draconis	Dra	Dragon
Equuleus	Equulei	Equ	Little Horse
Eridanus	Eridani	Eri	River Eridanus
Fornax	Fornacis	For	Furnace
Gemini	Geminorum	Gem	Twins
Grus	Gruis	Gru	Crane
Hercules	Herculis	Her	Hercules
Horologium	Horologii	Hor	Clock
Hydra	Hydrae	Hya	Water Snake
Hydrus	Hydri	Hyi	Lesser Water Snake
Indus	Indi	Ind	Indian
Lacerta	Lacertae	Lac	Lizard
Leo	Leonis	Leo	Lion
Leo Minor	Leonis Minoris	LMi	Little Lion
Lepus	Leporis	Lep	Hare
Libra	Librae	Lib	Scales
Lupus	Lupi	Lup	Wolf
Lynx	Lyncis	Lyn	Lynx
Lyra	Lyrae	Lyr	Lyre
Mensa	Mensae	Men	Table Mountain
Microscopium	Microscopii	Mic	Microscope
Monoceros	Monocerotis	Mon	Unicorn
Musca	Muscae	Mus	Fly
Norma	Normae	Nor	Set Square
Octans	Octantis	Oct	Octant
Ophiuchus	Ophiuchi	Oph	Serpent Bearer
Orion	Orionis	Ori	Orion
Pavo	Pavonis	Pav	Peacock
Pegasus	Pegasi	Peg	Pegasus
Perseus	Persei	Per	Perseus

Name	Genitive	Abbr.	English name
Phoenix	Phoenicis	Phe	Phoenix
Pictor	Pictoris	Pic	Painter's Easel
Pisces	Piscium	Psc	Fishes
Piscis Austrinus	Piscis Austrini	PsA	Southern Fish
Puppis	Puppis	Pup	Stern
Pyxis	Pyxidis	Pyx	Compass
Reticulum	Reticuli	Ret	Net
Sagitta	Sagittae	Sge	Arrow
Sagittarius	Sagittarii	Sgr	Archer
Scorpius	Scorpii	Sco	Scorpion
Sculptor	Sulptoris	Scu	Sculptor
Scutum	Scuti	Sct	Shield
Serpens	Serpentis	Ser	Serpent
Sextans	Sextantis	Sex	Sextant
Taurus	Tauri	Tau	Bull
Telescopium	Telescopii	Tel	Telescope
Triangulum	Trianguli	Tri	Triangle
Triangulum Australe	Trianguli Australis	TrA	Southern Triangle
Tucana	Tucanae	Tuc	Toucan
Ursa Major	Ursae Majoris	UMa	Great Bear
Ursa Minor	Ursae Minoris	UMi	Lesser Bear
Vela	Velorum	Vel	Sails
Virgo	Virginis	Vir	Virgin
Volans	Volantis	Vol	Flying Fish
Vulpecula	Vulpeculae	Vul	Fox

The Greek Alphabet

α	Alpha	ι	Iota	ρ	Rho
β	Beta	κ	Kappa	σ (ς)	Sigma
γ	Gamma	λ	Lambda	τ	Tau
δ	Delta	μ	Mu	υ	Upsilon
ε	Epsilon	ν	Nu	φ (φ)	Phi
ζ	Zeta	ξ	Xi	χ	Chi
η	Eta	ο	Omicron	ψ	Psi
θ	Theta	π	Pi	ω	Omega

Asterisms

Apart from the constellations (88 of which cover the whole sky), listed on pages 256–258, certain groups of stars, which may form a small part of a larger constellation, are readily recognizable and have been given individual names. These groups are known as *asterisms*, and the most famous (and well-known) is the Plough or Big Dipper, the common name for the seven brightest stars in the constellation of Ursa Major, the Great Bear. The names and details of some asterisms mentioned in this book are given in this list.

Some common asterisms

Belt of Orion	δ, ε and ζ Orionis
Big Dipper	α, β, γ, δ, ε, ζ and η Ursae Majoris
Cat's Eyes	λ and υ Scorpii
Circlet	γ, θ, ι, λ and κ Piscium
False Cross	ε and ι Carinae and δ and κ Velorum
Fish Hook	α, β, δ and π Scorpii
Guardians of the Pole	β and γ Ursae Minoris
Head of Cetus	α, γ, ξ², μ and λ Ceti
Head of Draco	β, γ, ξ and ν Draconis
Head of Hydra	δ, ε, ζ, η, ρ and σ Hydrae
Job's Coffin	α, β, γ and δ Delphini
Keystone	ε, ζ, η and π Herculis
Kids	ε, ζ and η Aurigae
Little Dipper	β, γ, η, ζ, ε, δ and α Ursae Minoris
Lozenge	β, γ, ξ and ν Draconis and ι Herculis
Milk Dipper	ζ, τ, σ, φ and λ Sagittarii
Plough	= Big Dipper
Pointers	α and β Ursae Majoris
Pot	= Venus Mirror
Venus' Mirror (or Saucepan)	ι, θ, ζ, ε, δ and η Orionis
Sickle	α, η, γ, ζ, μ and ε Leonis
Southern Pointers	α and β Centauri
Square of Pegasus	α, β and γ Pegasi with α Andromedae
Sword of Orion	θ and ι Orionis
Teapot	γ, ε, δ, λ, φ, σ, τ and ζ Sagittarii
Wain (or Charles' Wain)	= Big Dipper
Water Jar	γ, π, η and ζ Aquarii
Y of Aquarius	= Water Jar

107 Named stars brighter than magnitude 2.75

Name	Con	Mag	Name	Con	Mag
Achernar	α Eri	0.45	Aspidiske	ι Car	2.21
Acrab	β Sco	2.56	Athebyne	η Dra	2.73
Acrux	α Cru	0.77	Atria	α TrA	1.91
Adhara	ε CMa	1.50	Avior	ε Car	1.86
Aldebaran	α Tau	0.87	Bellatrix	γ Ori	1.64
Alderamin	α Cep	2.45	Betelgeuse	α Ori	0.45
Algieba	γ Leo	2.01	Canopus	α Car	−0.62
Algol	β Per	2.09	Capella	α Aur	0.08
Alhena	γ Gem	1.93	Caph	β Cas	2.28
Alioth	ε UMa	1.76	Castor	α Gem	1.58
Aljanah	ε Cyg	2.48	Deneb	α Cyg	1.25
Alkaid	η UMa	1.85	Denebola	β Leo	2.14
Almach	γ And	2.10	Diphda	β Cet	2.04
Alnair	α Gru	1.73	Dschubba	δ Sco	2.29
Alnilam	ε Ori	1.69	Dubhe	α UMa	1.81
Alnitak	ζ Ori	1.74	Elnath	β Tau	1.65
Alphard	α Hya	1.99	Eltanin	γ Dra	2.24
Alphecca	α CrB	2.22	Enif	ε Peg	2.38
Alpheratz	α And	2.07	Fomalhaut	α PsA	1.17
Alsephina	δ Vel	1.93	Gacrux	γ Cru	1.59
Altair	α Aql	0.76	Gienah	γ Crv	2.58
Aludra	η CMa	2.45	Hadar	β Cen	0.61
Ankaa	α Phe	2.40	Hamal	α Ari	2.01
Antares	α Sco	1.06	Hassaleh	ι Aur	2.69
Arcturus	α Boo	−0.05	Izar	ε Boo	2.35
Arneb	α Lep	2.58	Kaus Australis	ε Sgr	1.79
Ascella	ζ Sgr	2.60	Kaus Media	δ Sgr	2.72

Name	Con	Mag	Name	Con	Mag
Kochab	β UMi	2.07	**Procyon**	α CMi	0.40
Kraz	β Crv	2.65	**Rasalhague**	α Oph	2.08
Larawag	ε Sco	2.29	**Regulus**	α Leo	1.36
Lesath	υ Sco	2.70	**Rigel**	β Ori	0.18
Mahasim	θ Aur	2.65	**Rigil Kentaurus**	α Cen	−0.29
Markab	α Peg	2.49	**Ruchbah**	δ Cas	2.66
Markeb	κ Vel	2.47	**Sabik**	η Oph	2.43
Menkalinan	β Aur	1.90	**Sadr**	γ Cyg	2.23
Menkar	α Cet	2.54	**Saiph**	κ Ori	2.07
Menkent	θ Cen	2.06	**Sargas**	θ Sco	1.86
Merak	β UMa	2.34	**Scheat**	β Peg	2.44
Miaplacidus	β Car	1.67	**Schedar**	α Cas	2.24
Mimosa	β Cru	1.25	**Shaula**	λ Sco	1.62
Mintaka	δ Ori	2.25	**Sheratan**	β Ari	2.64
Mirach	β And	2.07	**Sirius**	α CMa	−1.44
Mirfak	α Per	1.79	**Spica**	α Vir	0.98
Mirzam	β CMa	1.98	**Suhail**	λ Vel	2.23
Mizar	ζ UMa	2.23	**Tarazed**	γ Aql	2.72
Muphrid	η Boo	2.68	**Tiaki**	β Gru	2.07
Naos	ζ Pup	2.21	**Unukalhai**	α Ser	2.63
Nunki	σ Sgr	2.05	**Vega**	α Lyr	0.03
Peacock	α Pav	1.94	**Wezen**	δ CMa	1.83
Phact	α Col	2.65	**Yed Prior**	δ Oph	2.73
Phecda	γ UMa	2.41	**Zosma**	δ Leo	2.56
Polaris	α UMi	1.97	**Zubenelgenubi**	α Lib	2.75
Pollux	β Gem	1.16	**Zubeneschamali**	β Lib	2.61
Porrima	γ Vir	2.74			

Further Information

Books

Bone, Neil (1993), *Observer's Handbook: Meteors*, George Philip, London & Sky Publ. Corp., Cambridge, Mass.

Chu, A (2012), *The Cambridge Photographic Moon Atlas*, Cambridge University Press, Cambridge

Dunlop, Storm (1999), *Wild Guide to the Night Sky*, HarperCollins, London

Dunlop, Storm (2012), *Practical Astronomy*, 3rd edn, Philip's, London

Dunlop, Storm, Rükl, Antonin & Tirion, Wil (2005), *Collins Atlas of the Night Sky*, HarperCollins, London

Heifetz, Milton & Tirion, Wil (2017), *A Walk through the Heavens*, 4th edn, Cambridge University Press, Cambridge

National Geographic & Wei-Haas, Maya (2023), *Stargazer's Atlas: The Ultimate Guide to the Night Sky*, National Geographic, Washington, D.C.

O'Meara, Stephen J. (2008), *Observing the Night Sky with Binoculars*, Cambridge University Press, Cambridge

Ridpath, Ian (2018), *Star Tales*, 2nd edn, Lutterworth Press, Cambridge

Ridpath, Ian, ed. (2003), *Oxford Dictionary of Astronomy*, 2nd edn, Oxford University Press, Oxford

Ridpath, Ian, ed. (2004), *Norton's Star Atlas*, 20th edn, Pi Press, New York

Ridpath, Ian & Tirion, Wil (2004), *Collins Gem – Stars*, HarperCollins, London

Ridpath, Ian & Tirion, Wil (2017), *Collins Pocket Guide Stars and Planets*, 5th edn, HarperCollins, London

Ridpath, Ian, McIntyre, Mary & Federman, Rachel (2024), *Collins Stargazer's Bible: Your Illustrated Companion to the Night Sky*, HarperCollins, London

Ridpath, Ian & Tirion, Wil (2019), *The Monthly Sky Guide*, 10th edn, Dover Publications, New York

Rükl, Antonín (1990), *Hamlyn Atlas of the Moon*, Hamlyn, London & Astro Media Inc., Milwaukee

Scagell, Robin (2000), *Philip's Stargazing with a Telescope*, George Philip, London

Stimac, Valerie (2019), *A Practical Guide to Astrotourism*, Lonely Planet, Franklin, TN

Tirion, Wil (2011), *Cambridge Star Atlas*, 4th edn, Cambridge University Press, Cambridge

Topalovic, Radmila & Kerss, Tom (2016), *Stargazing: Beginner's Guide to Astronomy*, HarperCollins, London

Journals

Astronomy, Astro Media Corp., 21027 Crossroads Circle, P.O. Box 1612, Waukesha, WI 53187-1612 USA. http://astronomy.com

Astronomy Now, Pole Star Publications, PO Box 175, Tonbridge, Kent TN10 4QX UK. http://astronomynow.com

Sky at Night Magazine, BBC publications, London. http://skyatnightmagazine.com

Sky & Telescope, Sky Publishing Corp., Cambridge, MA 02138-1200 USA. http://www.skyandtelescope.org/

Societies

American Association of Variable Star Observers (AAVSO), 49 Bay State Rd, Cambridge, MA 02138. Although primarily concerned with variable stars, the AAVSO also has a solar section. http://www.aavso.org/

American Astronomical Society (AAS), 1667 K Street NW, Suite 800, Washington, DC 20006, New York. http://aas.org/

American Meteor Society (AMS), Geneseo, New York. http://www.amsmeteors.org/

Association of Lunar and Planetary Observers (ALPO), ALPO Membership Secretary/Treasurer, P.O. Box 13456, Springfield, IL 62791-3456. An organization concerned with all forms of amateur astronomical observation, not just the Moon and planets, with numerous coordinated observing sections. http://alpo-astronomy.org/

Astronomical League (AL), 9201 Ward Parkway Suite #100, Kansas City, MO 64114. An umbrella organization consisting of over 240 local amateur astronomical societies across the United States. https://www.astroleague.org/

British Astronomical Association (BAA), Burlington House, Piccadilly, London W1J 0DU. http://www.britastro.org/ The principal British organization for amateur astronomers (with some professional members), particularly for those interested in carrying out observational programmes. Its membership is, however, worldwide. It publishes fully refereed, scientific papers and other material in its well-regarded journal.

Federation of Astronomical Societies, Secretary: John Stapleton http://www.fedastro.org.uk/fas/ An organization that is able to provide contact information for local astronomical societies in the United Kingdom.

Royal Astronomical Society, Burlington House, Piccadilly, London W1J 0BQ. http://www.ras.org.uk/ The premier astronomical society, with membership primarily drawn from professionals and experienced amateurs. It has an exceptional library and is a designated centre for the retention of certain classes of astronomical data. Its publications are the standard medium for dissemination of astronomical research.

Society for Popular Astronomy, 36 Fairway, Keyworth, Nottingham NG12 5DU. http://www.popastro.com/ A society for astronomical beginners of all ages, which concentrates on increasing members' understanding and enjoyment, but which does have some observational programmes. Its journal is entitled Popular Astronomy.

Software and Internet Sources

Planetary, Stellar and Lunar Visibility (planetary and eclipse freeware):
Alcyone Software, Germany.
http://www.alcyone.de

Redshift, Redshift Sky. https://redshiftsky.com

Starry Night & Starry Night Pro, Sienna Software Inc., Toronto, Canada.
http://www.starrynight.com

Stellarium, https://stellarium.org/

There are numerous sites with information about all aspects of astronomy, and all of those have numerous links. Although many amateur sites are excellent, treat any statements and data with caution. The sites listed below offer accurate information. Please note that the URLs may change. If so, use a good search engine, such as Google, to locate the information source.

Information

Astronomical data (inc. eclipses) HM Nautical Almanac Office:
http://astro.ukho.gov.uk

Astrophotography https://www.skyatnightmagazine.com/astrophotography/

Auroral information Michigan Tech:
http://www.geo.mtu.edu/weather/aurora/

Comets JPL Solar System Dynamics:
http://ssd.jpl.nasa.gov/

American Meteor Society:
http://amsmeteors.org/

Deep-sky objects Saguaro Astronomy Club Database:
http://www.virtualcolony.com/sac/

Eclipses NASA Eclipse Page:
http://eclipse.gsfc.nasa.gov/eclipse.html

Moon (inc. Atlas) Lunar and Planetary Institute (LPI):
https://www.lpi.usra.edu/resources/cla/

Planets Explore the planets:
https://science.nasa.gov/solar-system/planets/
Explore the Solar System: https://science.nasa.gov/solar-system/
Planetary Fact Sheets: http://nssdc.gsfc.nasa.gov/planetary/
planetfact.html

Satellites (inc. International Space Station)
Heavens Above: http://www.heavens-above.com/
Visual Satellite Observer: http://www.satobs.org/

Star Chart
> http://www.skyandtelescope.com/observing/interactive-sky-watching-tools/interactive-sky-chart/

What's Visible
> Skyhound: http://www.skyhound.com/sh/skyhound.html
> Skyview Cafe: http://www.skyviewcafe.com

Institutes and Organizations

European Space Agency: http://www.esa.int/

International Dark-Sky Association: http://www.darksky.org/

RASC Dark Sky: https://rasc.ca/

Jet Propulsion Laboratory: http://www.jpl.nasa.gov/

Lunar and Planetary Institute: http://www.lpi.usra.edu/

National Aeronautics and Space Administration: http://www.hq.nasa.gov/

Solar Data Analysis Center: http://umbra.gsfc.nasa.gov/

Space Telescope Science Institute: http://www.stsci.edu/

Acknowledgements

Our thanks to Ed Bloomer, Imo Bell, Catherine Muller and Sam Imperato, astronomers at Royal Observatory Greenwich.

Image Credits

Introduction diagrams pp. 6–33 Wil Tirion
p. 51 (left) CSNA/Siyu Zhang/Kevin M. Gill via Wikimedia Commons; (right) NASA via Wikimedia Commons
p. 55 ESA/Hubble via Wikimedia Commons
p. 63 Hubble/NASA/ESA via Wikimedia Commons
p. 67 NASA via Wikimedia Commons
p. 70 NASA
p. 75 (top) NASA via Wikimedia Commons; (bottom) NASA/ESA/A. van der Hoeven via Wikimedia Commons
p. 97 Hubble/ESA via Wikimedia Commons
p. 98 Jessie Eastlands via Wikimedia Commons
p. 105 (top) NASA; (bottom) ESO via Wikimedia Commons
p. 119 NASA
p. 127 (top) University of Tokyo; (bottom) ESO via Wikimedia Commons
p. 135 (top) Andrew Gray via Wikimedia Commons
p. 150 ESO/C. Malin via Wikimedia Commons
p. 152 Thomas K Vbg via Wikimedia Commons
p. 167 Adam Cuerden/NASA via Wikimedia Commons
p. 168 NASA via Wikimedia Commons
p. 169 NASA via Wikimedia Commons
p. 197 Judy Schmidt via Wikimedia Commons
p. 205 NASA/JPL via Wikimedia Commons
p. 213 Zelario 12 via Wikimedia Commons
p. 221 ESO/Y. Beletsky via Wikimedia Commons
p. 229 Kevin M. Gill via Wikimedia Commons
p. 231 EHT Collaboration via Wikimedia Commons

Index